RICHARD S. VARGA
Kent State University

Scientific Computation on Mathematical Problems and Conjectures

SOCIETY FOR INDUSTRIAL AND APPLIED MATHEMATICS
PHILADELPHIA, PENNSYLVANIA 1990

Printed by Capital City Press, Montpelier, Vermont.

Library of Congress Cataloging-in-Publication Data
Varga, Richard S.
Scientific computation on mathematical problems and conjectures / Richard S. Varga.
p. cm. — (CBMS–NSF regional conference series in applied mathematics; 60)
Includes bibliographical references.
ISBN 0-89871-257-2
1. Mathematics—Data processing. 2. Problem solving—Data processing. I. Title. II. Series.
QA76.95.V37 1990
510'.285—dc20 90-34761

Contents

Preface

The purpose of these lecture notes is to study in some detail the use of scientific computation as a tool in attacking a number of mathematical problems and conjectures. First, by scientific computation, we mean primarily computations which are carried out with a *large* number of significant digits, for calculations associated with a variety of numerical techniques, such as the (second) Remez algorithm in polynomial and rational approximation theory, Richardson extrapolation of sequences of numbers (to enhance convergence of these numbers), the accurate finding of zeros of polynomials of large degree, and the numerical approximation of integrals by quadrature techniques. We hasten to add that our goal here is not to delve into the specialized field dealing with the *creation* of robust and reliable software needed to implement these high-precision calculations, but rather to *emphasize* the enormous power that existing software brings to the mathematician's arsenal of weapons for attacking mathematical problems and conjectures.

As a case in point, we first study in Chapter 1 the Bernstein Conjecture of 1913 in polynomial approximation theory, which was recently settled (negatively) in 1985, by directly using high-precision calculations. In Chapter 2, we study the "1/9" Conjecture of 1977 in rational approximation theory. In this case, high-precision calculations gave strong indications that this conjecture was also false, but the final recent exact solution of this conjecture by A. A. Gonchar and E. A. Rakhmanov is a beautiful theoretical contribution, rather than a numerical one.

In Chapter 3, we briefly survey recent high-precision calculations, related to the famous Riemann Hypothesis of 1859, on finding zeros in the critical strip of the Riemann ζ-function. Then, we turn to the Pólya Conjecture of 1927, which is a weaker form of the Riemann Hypothesis. The affirmative analytic solution of the Pólya Conjecture, inspired by numerical computations, is given in this chapter. Next, results are given on recent high-precision numerical studies for determining lower bounds for the de Bruijn–Newman constant Λ. It turns out that this constant Λ satisfies $\Lambda \leq 0$ if the Riemann Hypothesis is true, and there is a complementary conjecture of 1976, of C. M. Newman, that states that $\Lambda \geq 0$.

In the remaining chapters, theoretical results related to three problems in analysis are dominant. The high-precision calculations in these chapters were of a different nature, in that either high-precision numerical experimentation was used to sharpen one's intuition for finding analytical results, or high-precision calculation of zeros of polynomials of large degree was used to obtain interesting graphical output.

As will be seen, the emphasis in this monograph rests strongly on the interplay between hard analysis and high-precision calculations. We note here that each chapter is independent of the remaining chapters, and each chapter has its own references given at the end of that chapter.

We gratefully acknowledge the comments and suggestions of my many friends and colleagues on the material presented here. At the lectures at Butler University, this author unwisely offered, as an inducement, one doughnut for each error found in the notes distributed for the CBMS–NSF Conference. (There were many times when I later wished that I had gone into the bakery business!) We are also indebted to Ms. Gail Bostic, of the Institute for Computational Mathematics at Kent State University, for her superb typing of this manuscript in LaTeX , to Gretchen M. Varga for her editing of the material in these pages, to Professor George Csordas for his valuable comments on Chapter 3, to Professor Arden Ruttan for having generated the interesting graphs in Chapter 5, and to Mr. Keith Fuller for having generated Figures 6.1 and 6.2 in Chapter 6.

Finally, above all, I wish to thank Professor Amos J. Carpenter for not only having so expertly executed all details of running this conference at Butler University, but also for the extreme care he took in reading the contents of this monograph, and for the figures of Chapters 1 and 4. As always, it was a pleasure to work with him.

CHAPTER 1

The Bernstein Conjecture in Approximation Theory

1.1. The Bernstein Conjecture.

For any real-valued function $f(x)$ defined on the interval $[-1,+1]$, its modulus of continuity is defined, for any $\delta > 0$, by

$$\omega(\delta; f) := \sup_{\substack{|x_1 - x_2| \le \delta \\ x_1, x_2 \in [-1,+1]}} |f(x_1) - f(x_2)|, \tag{1.1}$$

while its uniform norm on $[-1,+1]$ is defined by

$$\|f\|_{L_\infty[-1,+1]} := \sup\{|f(x)| : x \in [-1,+1]\}. \tag{1.1'}$$

With π_n denoting the set of all real polynomials of degree at most n $(n = 0, 1, \cdots)$, the following is a well-known result of Jackson [10] (cf. Meinardus [11, p. 56], Rivlin [14, p. 22]):

THEOREM 1. (Jackson [10]). *If $f(x)$ is a continuous real-valued function defined on $[-1,+1]$, then*

$$E_n(f) \le 6\omega\left(\frac{1}{n}; f\right) \qquad (n = 1, 2, \cdots), \tag{1.2}$$

where

$$E_n(f) := \inf\left\{\|f - g\|_{L_\infty[-1,+1]} : g \in \pi_n\right\}. \tag{1.3}$$

It is known (cf. [11, p. 16]) that, for any continuous real-valued function $f(x)$ defined on $[-1,+1]$, there exists a *unique* $\hat{p}_n(x) = \hat{p}_n(x; f)$ in π_n such that

$$E_n(f) = \|f - \hat{p}_n\|_{L_\infty[-1,+1]} \qquad (n = 0, 1, \cdots), \tag{1.3'}$$

and $\hat{p}_n(x)$ is called *the best uniform approximation* of $f(x)$ from π_n on $[-1,+1]$. Moreover, it is evident from (1.3) that the sequence $\{E_n(f)\}_{n=0}^{\infty}$ is a nonincreasing sequence of nonnegative numbers which, from the Weierstrass Approximation Theorem (cf. [14, p. 11]), tends to zero, i.e.,

$$\lim_{n\to\infty} E_n(f) = 0. \tag{1.4}$$

For the particular continuous function $|x|$ on $[-1,+1]$, it is easily seen that

$$\omega(\delta;|x|)=\delta \qquad (0<\delta\leq 1),$$

so that from (1.2) of Theorem 1,

$$E_n(|x|)\leq \frac{6}{n} \qquad (n=1,2,\cdots), \tag{1.4'}$$

which is a more precise form of (1.4) for the function $|x|$.

Since $|x|$ is an even continuous function on $[-1,+1]$, then so is its unique best uniform approximation from π_n on $[-1,+1]$, for any $n\geq 0$ (cf. Rivlin [14, p. 43, Exercise 1.1]). Combining this observation with the Chebyshev alternation characteristic of best uniform approximation by polynomials, it further follows (cf. [14, p. 26]) that

$$E_{2n}(|x|)=E_{2n+1}(|x|) \qquad (n=0,1,\cdots). \tag{1.5}$$

Therefore, for our purposes, it suffices to consider only the manner in which the sequence $\{E_{2n}(|x|)\}_{n=1}^{\infty}$ decreases to zero. From (1.4), we clearly have

$$2nE_{2n}(|x|)\leq 6 \qquad (n=1,2,\cdots). \tag{1.6}$$

To improve the upper bound of (1.6), Bernstein [2] suggested expanding $|x|$ in a Chebyshev series on $[-1,+1]$, i.e.,

$$|x|=\frac{4}{\pi}\left\{\frac{1}{2}+\sum_{m=1}^{\infty}\frac{(-1)^{m+1}T_{2m}(x)}{(2m-1)(2m+1)}\right\}, \tag{1.7}$$

where $T_n(x)$ denotes the nth Chebyshev polynomial (of the first kind). On taking the first n terms of the sum in (1.7), the absolute value of the remainder for (1.7) satisfies

$$\frac{4}{\pi}\left|\sum_{m=n+1}^{\infty}\frac{(-1)^{m+1}T_{2m}(x)}{(2m-1)(2m+1)}\right|\leq\frac{4}{\pi}\sum_{m=n+1}^{\infty}\frac{1}{(2m-1)(2m+1)},$$

since $|T_n(x)|\leq 1$ for all x in $[-1,+1]$. But, observing that

$$\frac{1}{(2m-1)(2m+1)}=\frac{1}{2}\left(\frac{1}{2m-1}-\frac{1}{2m+1}\right),$$

then this upper bound telescopes simply to

$$\begin{aligned}\frac{4}{\pi}\sum_{m=n+1}^{\infty}\frac{1}{(2m-1)(2m+1)} &= \frac{2}{\pi}\left\{\left(\frac{1}{2n+1}-\frac{1}{2n+3}\right)+\right.\\ &\qquad\left.\left(\frac{1}{2n+3}-\frac{1}{2n+5}\right)+\cdots\right\}\\ &= \frac{2}{\pi(2n+1)}.\end{aligned}$$

Thus,

$$\left| |x| - \frac{4}{\pi}\left\{\frac{1}{2} + \sum_{m=1}^{n} \frac{(-1)^{m+1}T_{2m}(x)}{(2m-1)(2m+1)}\right\}\right| \le \frac{2}{\pi(2n+1)}, \tag{1.8}$$

for all x in $[-1,+1]$ and all $n \ge 1$. Since the approximation to $|x|$ in (1.8) is a particular polynomial of degree $2n$, (1.8) implies that

$$2nE_{2n}(|x|) \le \frac{4n}{\pi(2n+1)} < \frac{2}{\pi} = 0.63661\cdots \qquad (n = 1, 2, \cdots), \tag{1.9}$$

which improves (1.6).

It turns out that the upper bound of $0.63661\cdots$ in (1.9) was significantly sharpened by Bernstein [2] in 1913. While (1.9) directly gives

$$\overline{\lim_{n\to\infty}}\ 2nE_{2n}(|x|) \le \frac{2}{\pi} = 0.63661\cdots, \tag{1.10}$$

Bernstein established, by means of a long and difficult proof, the following much deeper result:

THEOREM 2. (Bernstein [2]). *There exists a positive constant β (β for Bernstein) such that*

$$\lim_{n\to\infty} 2nE_{2n}(|x|) = \beta, \tag{1.11}$$

where β satisfies

$$0.278 < \beta < 0.286. \tag{1.12}$$

In addition to this above result, Bernstein noted in [2, p. 56], as a "curious coincidence," that the constant

$$\frac{1}{2\sqrt{\pi}} = 0.28209\ 47917\cdots \tag{1.13}$$

also satisfies the bounds of (1.12) and is very nearly the *average*, namely, 0.282, of the upper and lower bounds for β of (1.12). This observation has, over the years, become known as the

(1.14) **Bernstein Conjecture** (1913). $\beta \overset{?}{=} \dfrac{1}{2\sqrt{\pi}} = 0.28209\ 47917\cdots$.

In the more than 70 years since Bernstein's work appeared, the truth of this conjecture remained unresolved, despite numerical attacks by several authors (cf. Bell and Shah [1], Bojanic and Elkins [3], and Salvati [15]). The reasons that this conjecture remained open so long were probably due to the facts that (*i*) the accurate determination of the numbers $E_{2n}(|x|)$, for n large, is numerically *nontrivial*, and (*ii*) the convergence of $2nE_{2n}(|x|)$ to β, guaranteed by (1.11), is quite *slow*.

Recently, it was shown by Varga and Carpenter [16] in 1985 that the Bernstein Conjecture is *false*; this is a consequence of the following improved bounds of [16] for β (to be discussed in the next sections):

$$(1.15) \qquad 0.28016\ 85460\cdots = \ell_{20} \leq \beta \leq 2\mu_{100} = 0.28017\ 33791\cdots .$$

Since the upper bound for β in (1.15) is *less* than $1/(2\sqrt{\pi}) = 0.28209\ 47917\cdots$, the Bernstein Conjecture (1.14) is therefore *false*! In §1.2 below, we discuss high-precision calculations of $E_{2n}(|x|)$, while in §1.3 and §1.4, we briefly discuss calculations of [16] based on Bernstein's method for obtaining, respectively, upper and lower bounds for the Bernstein constant β. In the next section, §1.5, we discuss the Richardson extrapolation of the high-precision numbers $\{2nE_{2n}(|x|)\}_{n=1}^{52}$, which gives the following estimate for β to 50 significant figures:

$$(1.16) \qquad \beta \doteq \begin{array}{l} 0.28016\ 94990\ 23869\ 13303\ 64364 \\ \quad 91230\ 67200\ 00424\ 82139\ 81236, \end{array}$$

and in §1.6 we give a new conjecture on an asymptotic series expansion (6.3) for $2nE_{2n}(|x|)$. Finally, in §1.7 we include the related problem of best uniform *rational* approximation of $|x|$ on $[-1,+1]$.

1.2. Computing the numbers $\{2nE_{2n}(|x|)\}_{n=1}^{52}$ with high accuracy.

Let $\hat{p}_{2n}(x)$ in π_{2n} denote the unique best uniform approximation of $|x|$ from π_{2n} on $[-1,+1]$; i.e. (cf. (1.3′)),

$$(2.1) \qquad \||x| - \hat{p}_{2n}(x)\|_{L_\infty[-1,+1]} = E_{2n}(|x|) \quad (n = 1, 2, \cdots).$$

From the discussion in §1.1, because $|x|$ is even on $[-1,+1]$, then so is its best uniform approximation from π_{2n} on $[-1,+1]$, which implies that

$$(2.2) \qquad \hat{p}_{2n}(x) = \sum_{j=0}^{n} a_j(n)x^{2j} \qquad (n = 1, 2, \cdots).$$

We make the change of variables $x^2 = t, t \in [0,1]$, which changes our approximation problem to

$$(2.3)\ E_{2n}(|x|) = E_n\left(\sqrt{t}; [0,1]\right) := \inf\left\{\|\sqrt{t} - h_n(t)\|_{L_\infty[0,1]} : h_n \in \pi_n\right\}.$$

If we write

$$(2.4) \qquad E_n\left(\sqrt{t}; [0,1]\right) = \|\sqrt{t} - \hat{h}_n(t)\|_{L_\infty[0,1]} \qquad (\hat{h}_n \in \pi_n),$$

then clearly (cf. (2.2))

$$(2.5) \qquad \hat{p}_{2n}(x) = \hat{h}_n(x^2) \qquad (n = 1, 2, \cdots).$$

Thus, determining the values $E_{2n}(|x|)$ and $\hat{p}_{2n}(x)$ is equivalent to determining $E_n(\sqrt{t}; [0,1])$ and $\hat{h}_n(t)$.

The minimization problem (2.3) was solved using the following essentially standard implementation of the (second) Remez algorithm (cf. Meinardus [11, p. 105]):

Step 1. Let $S := \{t_j\}_{j=0}^{n+1}$ be a set of $n+2$ distinct points in $[0,1]$ satisfying

$$0 \le t_0 < t_1 < \cdots < t_{n+1} \le 1. \tag{2.6}$$

Step 2. Find the unique polynomial $h_n(t)$ and the constant λ (which is a linear problem) such that

$$h_n(t_j) + (-1)^j \lambda = \sqrt{t_j} \qquad (j = 0, 1, \cdots, n+1). \tag{2.7}$$

Thus, $h_n(t)$ is the best uniform approximation from π_n to $\sqrt{t}$ on this discrete set S, with an alternating error $|\lambda|$ in successive points t_j of S. Thus, in analogy with the notation of (2.3), we can write

$$\|\sqrt{t} - h_n(t)\|_{L_\infty(S)} = E_n(\sqrt{t}; S) = |\lambda|. \tag{2.8}$$

Because S is a subset of $[0,1]$, then clearly

$$\|\sqrt{t} - h_n(t)\|_{L_\infty[0,1]} - |\lambda| \ge 0. \tag{2.9}$$

Step 3. With a preassigned (small) $\varepsilon > 0$, if $\|\sqrt{t} - h_n(t)\|_{L_\infty[0,1]} - |\lambda| \le \varepsilon$, the iteration is terminated. Otherwise, find a new set S' from the set of local extrema in $[0,1]$, with alternating signs, of the function $\sqrt{t} - h_n(t)$ from the previous Step 2, and *repeat* Steps 2 and 3 until the termination criterion is satisfied.

From a computational point of view, it is useful to know that the sequence of $|\lambda|$'s generated from repeated applications of this Remez algorithm is monotone *increasing*.

Starting with the particular alternation set $S^{(0)} := \{t_j^{(0)}\}_{j=0}^{n+1}$, where

$$t_j^{(0)} := \frac{1}{2}\left\{1 + \cos\left[\frac{(n+1-j)\pi}{n+1}\right]\right\} \qquad (j = 0, 1, \cdots, n+1) \tag{2.10}$$

are the $n+2$ extreme points of the Chebyshev polynomial $T_{n+1}(2t-1)$ on $[0,1]$, and using Brent's MP package [4] to handle the multiple-precision computations on a VAX 11/780 in the Department of Mathematical Sciences at Kent State University, the iterates of the Remez algorithm were terminated when $\|\sqrt{t} - h_n(t)\|_{L_\infty[0,1]}$ and $|\lambda|$ agreed (cf. (2.9)) to 100 significant digits. Because of the known quadratic convergence of this (second) Remez algorithm (cf. [11, p. 113]), at most *nine* iterations were needed for convergence in each case considered. Taking into account guard digits and the possibility of some small rounding errors, we believe that the numbers $\{E_{2n}(|x|)\}_{n=1}^{52}$ we determined are accurate to at least 95 significant digits.

To conserve space, we give the products $\{2nE_{2n}(|x|)\}_{n=1}^{52}$, truncated to 20 significant digits in Table 1.1, to show the slow convergence of this sequence. (Printouts of $\{2nE_{2n}(|x|)\}_{n=1}^{52}$ to 100 significant digits are available on request.)

TABLE 1.1
$\{2nE_{2n}(|x|)\}_{n=1}^{52}$.

n	$2nE_{2n}(\|x\|)$	n	$2nE_{2n}(\|x\|)$
1	0.25000 00000 00000 00000	27	0.28010 92365 22206 18525
2	0.27048 35971 11137 10107	28	0.28011 34608 89950 28384
3	0.27557 43724 01175 38604	29	0.28011 72562 49499 61792
4	0.27751 78246 75052 69646	30	0.28012 06787 72662 82833
5	0.27845 11855 35508 60152	31	0.28012 37757 31660 88450
6	0.27896 79174 64958 70636	32	0.28012 65871 38731 91844
7	0.27928 29449 58518 02460	33	0.28012 91470 43904 51720
8	0.27948 88375 94507 44771	34	0.28013 14845 70012 61069
9	0.27963 06574 10128 20125	35	0.28013 36247 44030 04676
10	0.27973 24337 71973 82968	36	0.28013 55891 69271 11713
11	0.27980 79172 88743 87383	37	0.28013 73965 72336 69662
12	0.27986 54321 23793 27279	38	0.28013 90632 50782 89591
13	0.27991 02543 15557 69036	39	0.28014 06034 41582 48218
14	0.27994 58584 85782 13247	40	0.28014 20296 25997 94087
15	0.27997 46066 86407 49231	41	0.28014 33527 83104 08169
16	0.27999 81519 56316 72827	42	0.28014 45826 01611 08707
17	0.28001 76771 33297 25379	43	0.28014 57276 57645 50097
18	0.28003 40474 14993 50964	44	0.28014 67955 64600 41624
19	0.28004 79072 85905 85156	45	0.28014 77930 99959 13546
20	0.28005 97447 60423 15265	46	0.28014 87263 13048 74446
21	0.28006 99348 31809 43067	47	0.28014 96006 16931 43684
22	0.28007 87694 75287 53423	48	0.28015 04208 67046 95023
23	0.28008 64787 57075 57049	49	0.28015 11914 28744 92326
24	0.28009 32459 38808 50547	50	0.28015 19162 35465 27355
25	0.28009 92184 52382 83558	51	0.28015 25988 39017 81632
26	0.28010 45159 86556 70489	52	0.28015 32424 53163 84249

It appears that the products $\{2nE_{2n}(|x|)\}_{n=1}^{52}$ in Table 1.1 have converged to four significant digits. Based on the asymptotic estimates of §1.6, in order to have $|2nE_{2n}(|x|) - \beta| < 10^{-10}$, one would need $n \geq 20,968$. This would require finding polynomials of best uniform approximation to $\sqrt{t}$ on $[0,1]$ of degree at least 10,484, which would be a truly formidable computation!

1.3. Computing upper bounds for the Bernstein constant β.

To obtain upper and lower bounds for the Bernstein constant β, Bernstein [2] introduced the particular function

$$F(t) := \sum_{k=0}^{\infty} \frac{t}{(t+2k+1)^2 - \frac{1}{4}} \qquad (t \geq 0). \tag{3.1}$$

With the psi (digamma) function $\Psi(z)$ (cf. Whittaker and Watson [19, p.

240]), defined by

$$\Psi(z) := \frac{d}{dz}(\log \Gamma(z)) = \frac{\Gamma'(z)}{\Gamma(z)}, \tag{3.2}$$

the function $F(t)$ of (3.1) has the representation

$$F(t) = \frac{t}{2}\left\{\Psi\left(\frac{t}{2}+\frac{3}{4}\right) - \Psi\left(\frac{t}{2}+\frac{1}{4}\right)\right\} \qquad (t \geq 0). \tag{3.3}$$

Other representations for $F(t)$ (cf. [2]) include

$$F(t) = \frac{t}{2t+1}F\left(1,1;t+\frac{3}{2};\frac{1}{2}\right) = t\int_0^1 \frac{z^{t-1/2}dz}{z+1} = \frac{1}{2}\int_0^\infty \frac{e^{-u}du}{\cosh(u/2t)}, \tag{3.4}$$

where $F(a,b;c;z)$ denotes the classical hypergeometric function (cf. Henrici [9, p. 56]). The last integral in (3.4) shows that $F(t)$ is strictly increasing on $[0,+\infty)$, with $F(0)=0$ and $F(+\infty)=\frac{1}{2}$.

At this point, it is not at all obvious *why* the function $F(t)$ of (3.1) should play a role in the determination of the Bernstein constant β. To indicate *how* this function arises, Bernstein considered the following polynomial interpolation problem. For each fixed positive integer n, consider the specific $2n+1$ distinct points in $[-1,+1]$ given by

$$x_0 := 0; x_k := \cos\left[\frac{(k-1/2)\pi}{2n}\right] \qquad (k=1,2,\cdots,2n). \tag{3.5}$$

Since the points $\{x_k\}_{k=1}^{2n}$ are just the zeros of the Chebyshev polynomial $T_{2n}(x)$, we note that

$$\omega(x) := \prod_{j=0}^{2n}(x-x_j) = xT_{2n}(x). \tag{3.6}$$

If $R_{2n}(x)$ denotes the unique polynomial in π_{2n} which interpolates $|x|$ in the $2n+1$ points of (3.5), then $R_{2n}(x)$ is an even polynomial in x which, after some manipulations, satisfies

$$|x| - R_{2n}(x) = \frac{T_{2n}(x)}{n}H_{2n}(x) \qquad (x \in [0,1]), \tag{3.7}$$

where

$$H_{2n}(x) := -\sum_{k=0}^{n-1}\frac{(-1)^k \sin[(k+\frac{1}{2})\frac{\pi}{2n}]}{x+\cos[(k+\frac{1}{2})\frac{\pi}{2n}]} \qquad (x\in[0,1]). \tag{3.7'}$$

By means of a long proof, Bernstein [2] showed that

$$|x| - R_{2n}(x) = \frac{T_{2n}(x)}{n}\left[F\left(\frac{2nx}{\pi}\right)+\eta_n(x)\right] \qquad (x\in[0,1]), \tag{3.8}$$

where

$$|\eta_n(x)| \leq \frac{4+\pi^2}{2n^{2/5}} \qquad (x\in[0,1]; n=1,2,\cdots). \tag{3.9}$$

Next, by definition,

$$E_{2n}(|x|) \le \| |x| - R_{2n}(x) \|_{L_\infty[-1,+1]} = \| |x| - R_{2n}(x) \|_{L_\infty[0,1]},$$

the last equality arising from the fact that $|x|$ and $R_{2n}(x)$ are both even functions. Then, (3.8) and (3.9) imply, since $|T_{2n}(x)| \le 1$ in $[0,1]$ and since $F(t)$ is increasing on $[0,+\infty)$ with $F(+\infty) = \frac{1}{2}$, that

$$\overline{\lim_{n\to\infty}}\ 2nE_{2n}(|x|) \le 2F(+\infty) = 1. \tag{3.10}$$

Note that (3.10), while improving upon (1.6), is not as good as the result of (1.10). Moreover, (3.8) shows that the errors, $|x| - R_{2n}(x)$, are *far* from equioscillating in $[-1,+1]$ for n large, because the strictly increasing nature of $F(t)$ implies that the largest such errors can occur only in the neighborhoods of $x = \pm 1$, and none can occur in the neighborhood of $x = 0$.

To introduce equioscillations near $x = 0$, Bernstein [2] suggested the following. For n any *even* integer, first note that

$$T_{2n}(x) = \cos(2n \arccos x) = \cos(2n \arcsin x). \tag{3.11}$$

Next, for a *fixed* nonnegative integer m, as well as for all $n > m$, let $\{\xi_k(2n) := \sin[(2k-1)\pi/4n]\}_{k=1}^m$ denote the m smallest zeros of $T_{2n}(x)$ in $[0,1]$. Clearly, $T_{2n}(x)/\left(x^2 - \xi_k^2(2n)\right)$ is an even polynomial in π_{2n} for each k with $1 \le k \le m$, so that the polynomial $Q_{2n}(x)$, defined by

$$Q_{2n}(x) := R_{2n}(x) + \frac{T_{2n}(x)}{n}\left\{a_0 + \left(\frac{\pi}{2n}\right)^2 \sum_{k=1}^m \frac{a_k}{x^2 - \xi_k^2(2n)}\right\}, \tag{3.12}$$

is also an even polynomial in π_{2n} for each $n > m$. In addition, from (3.8) and (3.12), it follows with $x = \pi b/(2n)$ that

(3.13)

$$|x| - Q_{2n}(x) = \frac{T_{2n}(x)}{n}\left\{F(b) - \left(a_0 + \sum_{k=1}^m \frac{a_k}{b^2 - [\frac{2n}{\pi}\sin(\frac{(2k-1)\pi}{4n})]^2}\right) + \eta_n(x)\right\},$$

for every $n > m$. Define now

(3.14)

$$\mu_m := \inf_{\substack{a_0, a_1, \cdots, a_m \\ \text{real}}} \left\{ \left\| \cos(\pi b)\left[F(b) - \left(a_0 + \sum_{k=1}^m \frac{a_k}{b^2 - (\frac{2k-1}{2})^2}\right)\right]\right\|_{L_\infty[0,+\infty)}\right\},$$

for each nonnegative integer m. But since from (3.11), for a fixed $b \ge 0$,

$$T_{2n}(x) = \cos\left[2n \arcsin\left(\frac{\pi b}{2n}\right)\right] \to \cos(\pi b) \qquad (n \to \infty),$$

it can be verified, in analogy with (3.10), that

$$\overline{\lim_{n\to\infty}}\ 2nE_{2n}(|x|) \le 2\mu_m \qquad (m = 0, 1, \cdots). \tag{3.15}$$

The sequence of positive constants $\{\mu_m\}_{m=0}^{\infty}$ from (3.14) is clearly nonincreasing:

$$\mu_0 \geq \mu_1 \geq \mu_2 \geq \cdots , \tag{3.16}$$

and thus convergent. Now, Bernstein [2, p. 55] proved that the Bernstein constant β of (1.11) and the limit of the sequence $\{\mu_m\}_{m=0}^{\infty}$ are connected through

$$\beta = 2 \lim_{m\to\infty} \mu_m. \tag{3.17}$$

Thus with (3.16), each constant μ_m of the approximation problem (3.14) provides the following *upper bound* for β:

$$\beta \leq 2\mu_m \qquad (m = 0, 1, \cdots). \tag{3.18}$$

Interestingly, Bernstein [2] numerically estimated in 1913 the solution of (3.14) for $m = 3$ and found that $\mu_3 < 0.143$, so that

$$2\mu_3 < 0.286,$$

which is the upper bound for β mentioned in (1.12). (More accurate estimates of $2\mu_m$ can be found in Table 1.2 below.)

TABLE 1.2
$\{2\mu_m\}$.

m	$2\mu_m$	m	$2\mu_m$
0	0.50000 00000 00000	10	0.28056 81480 84662
1	0.30981 66482 77486	20	0.28026 79181 28026
2	0.28964 46428 36759	30	0.28021 30013 47551
3	0.28458 56232 64382	40	0.28019 38951 81171
4	0.28268 16444 08752	50	0.28018 50827 23738
5	0.28177 99926 24272	60	0.28018 03067 66681
6	0.28128 65208 69723	70	0.28017 74317 42434
7	0.28098 84334 65837	80	0.28017 55680 33390
8	0.28079 50582 78109	90	0.28017 42915 00582
9	0.28066 26720 87176	100	0.28017 33791 01718

Now, the minimization problem in (3.14), to determine the constant μ_m, appears to be special weighted rational approximation of $F(b)$ on $[0, +\infty)$, but the weight, $\cos(\pi b)$, in (3.14) is certainly not of one sign on this interval. But, as pointed out in [16], it turns out that the solution of the approximation problem in (3.14) has an interesting oscillation character that permits the use of a modified form of the (second) Remez algorithm. (It should be emphasized that the work of Bernstein [2] in 1913 *predates* the 1934 appearance of Remez's algorithm [13]!) The minimization problem (3.14) was solved using the following modified (second) Remez algorithm:

Step 1. Let $\tilde{S} := \{t_j\}_{j=0}^{m+1} \quad (m \geq 1)$ be a set of $m + 2$ distinct points in the interval $[0, +\infty]$ satisfying

$$0 = t_0 < t_1 < \cdots < t_m \leq m - \frac{1}{2} < t_{m+1} =: +\infty. \tag{3.19}$$

Step 2. Find the $m+2$ unique constants $\{a_i\}_{i=0}^m$ and λ (which is a linear problem), such that

$$(3.20)\quad \begin{cases} \cos(\pi t_j)\left\{a_0 + \sum_{k=1}^m \frac{a_k}{t_j^2 - [(2k-1)/2]^2}\right\} \\ \qquad\qquad +(-1)^{j+1}\lambda = \cos(\pi t_j)F(t_j) \quad (j=0,\cdots,m), \\ a_0 + \lambda = \frac{1}{2} = F(+\infty). \end{cases}$$

The solution of the linear problem (3.20) forces the function

$$(3.21)\qquad R_m(t) := \cos(\pi t)\left[F(t) - \left(a_0 + \sum_{k=1}^m \frac{a_k}{t^2 - [(2k-1)/2]^2}\right)\right]$$

to equioscillate on the subset $\{t_j\}_{j=0}^m$ of $\tilde{S}$ with an alternating error $|\lambda|$ in the successive points of $\{t_j\}_{j=0}^m$, and, in addition, forces $R_m(t)$ to oscillate between $+|\lambda|$ and $-|\lambda|$ as $t \to \infty$ (see Figure 1.1). Thus, in analogy with (2.8), we have

$$(3.22)\qquad \|R_m(t)\|_{L_\infty(\tilde{S})} = |\lambda|,$$

and, because $\tilde{S}$ is a subset of $[0,+\infty]$, then (cf. (2.9))

$$(3.23)\qquad \|R_m(t)\|_{L_\infty[0,+\infty)} - |\lambda| \geq 0.$$

As a background for the conditions of the next step of this modified Remez algorithm, we observe that

$$(3.24)\qquad \begin{aligned} &\frac{d}{dt}\left\{F(t) - \left(a_0 + \sum_{k=1}^m \frac{a_k}{t^2 - [(2k-1)/2]^2}\right)\right\} \\ &\qquad = F'(t) + 2t\sum_{k=1}^m \frac{a_k}{[t^2 - [(2k-1)/2]^2]^2}. \end{aligned}$$

Thus, if the solution of (3.20) is such that the a_k's $(0 \leq k \leq m)$ and λ are all positive numbers, then the strictly increasing character of $F(t)$ gives, with (3.24), that

$$G_m(t) := F(t) - \left(a_0 + \sum_{k=1}^m \frac{a_k}{t^2 - [(2k-1)/2]^2}\right)$$

is strictly increasing from $-\infty$ to $+\lambda$ on the interval $(m - \frac{1}{2}, +\infty)$. Defining

$\tau_m = \tau_m(\tilde{S})$ to be the unique value of t in $(m - \frac{1}{2}, +\infty)$ such that $G_m(\tau_m) = -\lambda$, then because $|R_m(t)| = |\cos(\pi t) \cdot G_m(t)| \leq |G_m(t)| \leq \lambda$ for all $t \geq \tau_m$, and because $G_m(t) \to \frac{1}{2} - a_0 = \lambda$ as $t \to +\infty$, it follows that

$$\|R_m(t)\|_{L_\infty[\tau_m, +\infty)} = |\lambda|.$$

On the other hand, since $\{t_j\}_{j=0}^m$ is a subset of $[0, m - \frac{1}{2}]$, then from (3.22), $\|R_m(t)\|_{L_\infty[0,m-1/2]} \geq |\lambda|$. Hence, because $[0, m - \frac{1}{2}]$ is a subset of $[0, \tau_m]$,

$$\|R_m(t)\|_{L_\infty[0,\tau_m(\tilde{S})]} = \|R_m(t)\|_{L_\infty[0,+\infty)}, \tag{3.25}$$

and it can then be numerically determined whether or not the stronger statement

$$\|R_m(t)\|_{L_\infty[0,m-1/2]} = \|R_m(t)\|_{L_\infty[0,+\infty)} \tag{3.25'}$$

is valid. (In the special cases treated thus far in the literature for determining upper bounds for β, such as $m = 1$ of Bojanic and Elkins [3] and $m = 3$ of Bernstein [2], the choices of the associated sets $\tilde{S}$ *were* such that the a_k's $(0 \leq k \leq m)$ and λ of Step 2 were positive, and (3.25′) was also satisfied.) This brings us to Step 3.

Step 3. With a preassigned (small) $\varepsilon > 0$, if the solutions $\{a_k\}_{k=0}^m$ and λ of Step 2 are all positive numbers, if (3.25′) is satisfied, and if

$$\|R_m(t)\|_{L_\infty[0,m-1/2]} - \lambda \leq \varepsilon,$$

the iteration is terminated. Otherwise, find a new set $\tilde{S}'$ consisting of a set of m local extrema of $R_m(t)$ (with alternating signs) in $(0, m - \frac{1}{2}]$, in addition to the points $t_0 := 0$ and $t_{m+1} := \infty$, and *repeat* Steps 2 and 3 until the termination criterion is satisfied.

Starting with the particular set $\tilde{S}^{(0)} := \{t_j^{(0)}\}_{j=0}^{m+1}$, where

$$t_0^{(0)} := 0; \quad t_j^{(0)} := \frac{2j-1}{2} \quad (j = 1, 2, \cdots, m); \quad t_{m+1}^{(0)} := +\infty, \tag{3.26}$$

the associated system of linear equations (3.20) was solved for $\{a_k^{(0)}\}_{k=0}^m$ and $\lambda^{(0)}$, using Gaussian elimination with partial pivoting. In every case considered in the tabulation in Table 1.2, the starting values (3.26) for the set $\tilde{S}$ of alternation points were sufficiently good so that Steps 2 and 3 of the above modified Remez algorithm always produced positive numbers $\{a_k\}_{k=0}^m$ and λ, as well as alternation sets satisfying (3.19) and (3.25′). Moreover, the convergence of this algorithm was, as might be expected, *quadratic*, and at most 10 iterations of this modified Remez algorithm were needed for convergence in the cases considered.

In Figure 1.1, we graph the function $R_5(t)$ of (3.21), which is associated with the best uniform approximation constant μ_5 of (3.14). In this figure, there are six alternation points (denoted by small dark disks) in the interval $[0, 9/2]$, as well as oscillations in $(9/2, +\infty)$ that grow in modulus to μ_5 as $t \to \infty$.

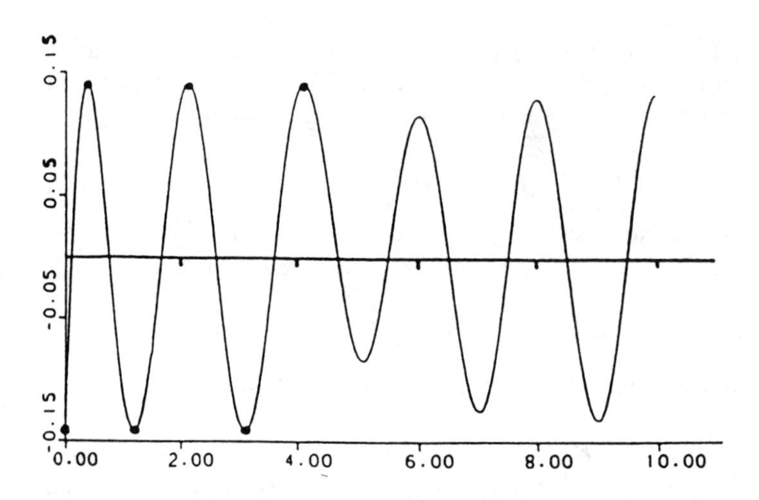

Figure 1.1: $R_5(t)$.

Unfortunately, the above calculations were *not* carried out to the same high accuracy (100 significant digits) as were the calculations of $\{2nE_{2n}(|x|)\}_{n=1}^{52}$ of §1.2. One reason for this is that the modified Remez algorithm applied to the minimization problem of (3.14) necessarily requires repeated evaluation of the function $F(t)$ of (3.1). Here, we used the representation (3.3) of $F(t)$ in terms of the psi function $\Psi(x)$, and $\Psi(x)$ was approximated, based on the work of Cody, Strecok, and Thacher [7], by

$$\Psi(x) \doteq (x - x_0) r_{8,8}(x) \qquad \left(\frac{1}{2} \le x \le 3\right), \tag{3.27}$$

where $x_0 = 1.46163\cdots$ is the unique positive zero of $\Psi(x)$ known to 40 significant digits, and where (from [7]) $r_{8,8}(x)$ is the ratio of two specific polynomials in x of degree 8; and by

$$\Psi(x) \doteq \ln x - \frac{1}{2x} + r_{6,6}\left(\frac{1}{x^2}\right) \qquad (3 \le x < \infty), \tag{3.28}$$

where (from [7]) $r_{6,6}(u)$ is the ratio of two specific polynomials in u of degree 6. For the range $0 < x \le \frac{1}{2}$, we used the known recurrence relation for the function $\Psi(x)$

$$\Psi(x) = \Psi(x+1) - \frac{1}{x}, \tag{3.29}$$

and the approximation of (3.27).

Now, the above approximations of $\Psi(t)$ are good to about 20 significant digits (cf. [7]), so we estimate that our calculations of $\{2\mu_m\}_{m=0}^{100}$ are accurate to at least 15 significant digits. To conserve space, a subset of the numbers $\{2\mu_m\}_{m=0}^{100}$ is given, truncated to 15 significant digits in Table 1.2.

It is evident from Table 1.2 that from (1.13) and (3.16)–(3.17),

$$\frac{1}{2\sqrt{\pi}} = 0.28209\,47917\cdots > 2\mu_5 > \beta, \tag{3.30}$$

so that the Bernstein Conjecture is *false*!

1.4. Computing lower bounds for the Bernstein constant β.

We have seen that our calculations in §1.3 of the upper bounds $\{2\mu_m\}_{m=0}^{100}$ of the Bernstein constant β (cf. (1.11)) are *sufficient* to disprove Bernstein's Conjecture (1.14). Thus, in terms of settling Bernstein's Conjecture, it is obviously *unnecessary* to determine lower bounds for β. However, to round out our discussion, we include here calculations of lower bounds for β based on another ingenious method of Bernstein. The calculations of these lower bounds for β proved to be the most *laborious* of all the calculations we performed.

To describe Bernstein's method [2] of determining lower bounds for β, we define

$$\phi_m(x) := \prod_{j=1}^{m-1} (x^2 - j^2) \qquad (m \geq 1), \tag{4.1}$$

and

$$\psi_m(x) = \psi_m(x; \lambda_1, \lambda_2, \cdots, \lambda_m) := \prod_{j=1}^{m} (x^2 - \lambda_j^2) \quad (m \geq 1). \tag{4.2}$$

(Here, we use the convention that $\prod_{j=\beta}^{\alpha} := 1$ if $\alpha < \beta$.) The parameters λ_j appearing in (4.2) are assumed to satisfy

$$j - 1 < \lambda_j < j \qquad (j \geq 1). \tag{4.3}$$

Then, for each $m \geq 1$, set

$$B_m(\lambda_1, \lambda_2, \cdots, \lambda_m) := \frac{\displaystyle\sum_{i=1}^{m} \frac{\phi_m(\lambda_i)}{\psi_m'(\lambda_i)} \left[1 - \left(\frac{2\lambda_i}{\lambda_i + \frac{1}{2}}\right) F\left(\lambda_i + \frac{1}{2}\right)\right]}{\displaystyle\sum_{i=1}^{m} \frac{\phi_m(\lambda_i)}{\psi_m'(\lambda_i)} \left[\frac{2}{\pi \lambda_i} + \tan\left(\frac{\pi}{2}[\lambda_i - i + 1]\right)\right]}, \tag{4.4}$$

where the function $F(t)$ is defined in (3.1). Note that from (4.1) and (4.2), we can write

$$\frac{\phi_m(\lambda_i)}{\psi_m'(\lambda_i)} = \frac{\displaystyle\prod_{j=1}^{i-1} (\lambda_i^2 - j^2) \cdot \prod_{j=i}^{m-1} (j^2 - \lambda_i^2)}{\displaystyle 2\lambda_i \prod_{j=1}^{i-1} (\lambda_i^2 - \lambda_j^2) \cdot \prod_{j=i+1}^{m} (\lambda_j^2 - \lambda_i^2)} \qquad (1 \leq i \leq m). \tag{4.5}$$

With the conditions of (4.3), we see that the above ratios are all positive. Thus, since $F(t)$ is strictly increasing on $[0, +\infty)$ with $F(0) = 0$ and with

$F(+\infty) = \frac{1}{2}$ (cf. §1.3), we deduce that each term of either sum of (4.4) is necessarily positive, whence $B_m(\lambda_1, \lambda_2, \cdots, \lambda_m) > 0$.

With β as defined in (1.11), Bernstein [2] showed that $B_m(\lambda_1, \lambda_2, \cdots, \lambda_m)$ is a lower bound for β, i.e.,

$$\beta \geq B_m(\lambda_1, \lambda_2, \cdots, \lambda_m), \tag{4.6}$$

for each positive integer m and for each choice of $\{\lambda_j\}_{j=1}^m$ satisfying (4.3). The best such lower bound for β for each $m \geq 1$ is clearly given by

$$\ell_m := \max\left\{B_m(\lambda_1, \lambda_2, \cdots, \lambda_m) : \{\lambda_j\}_{j=1}^m \text{ satisfies } (4.3)\right\}, \tag{4.7}$$

so that

$$\beta \geq \ell_m > 0 \qquad (m \geq 1). \tag{4.8}$$

Next, consider any parameters $\{\lambda_j\}_{j=1}^{m+1}$ satisfying (4.3). On fixing $\{\lambda_j\}_{j=1}^m$ and on letting λ_{m+1} decrease to m, it can readily be verified from (4.4) and (4.5) that

$$\lim_{\lambda_{m+1} \to m} B_{m+1}(\lambda_1, \lambda_2, \cdots, \lambda_m, \lambda_{m+1}) = B_m(\lambda_1, \lambda_2, \cdots, \lambda_m). \tag{4.9}$$

As a consequence, we see from (4.7) that

$$\ell_{m+1} \geq \ell_m \qquad (m \geq 1), \tag{4.10}$$

so that, with (4.8), $\{\ell_m\}_{m=1}^\infty$ is a bounded nondecreasing sequence of positive numbers. Now, Bernstein [2] further showed that the limit of this sequence is β:

$$\beta = \lim_{m \to \infty} \ell_m. \tag{4.11}$$

Bernstein in fact numerically estimated ℓ_1 and ℓ_2 in [2] and found that

$$\ell_1 > 0.27 \quad \text{and} \quad \ell_2 > 0.278. \tag{4.12}$$

This last estimate of ℓ_2 appears as the lower bound of β in (1.12). (More accurate estimates of ℓ_1 and ℓ_2 can be found in Table 1.3.)

TABLE 1.3
$\{\ell_m\}_{m=1}^{20}$.

m	ℓ_m	m	ℓ_m
1	0.27198 23590 30477	11	0.28016 34641 87524
2	0.27893 09228 49406	12	0.28016 48933 27009
3	0.27981 10004 37231	13	0.28016 59052 38063
4	0.28002 43339 28903	14	0.28016 66415 27680
5	0.28009 77913 15214	15	0.28016 71898 92928
6	0.28012 91830 79687	16	0.28016 76066 00825
7	0.28014 46910 09336	17	0.28016 79288 71653
8	0.28015 31877 11753	18	0.28016 81819 90114
9	0.28015 82176 99044	19	0.28016 83835 39180
10	0.28016 13794 71687	20	0.28016 85460 02042

We now describe our calculations of the lower bounds ℓ_m. It is evident from (4.4) that the parameters $\lambda_1, \lambda_2, \cdots, \lambda_m$ enter nonlinearly in the definition of $B_m(\lambda_1, \lambda_2, \cdots, \lambda_m)$. Computationally, we used a fairly standard optimization (maximization) routine, without derivatives, to determine the real parameters $\{\lambda_i\}_{i=1}^m$, subject to the constraints of (4.3), which maximized $B_m(\lambda_1, \lambda_2, \cdots, \lambda_m)$, thereby determining ℓ_m of (4.7). Again, because the function $F(t)$ appears explicitly in the definition of $B_m(\lambda_1, \lambda_2, \cdots, \lambda_m)$ of (4.4), we used the approximations of (3.27)–(3.29) for the psi function $\Psi(t)$ and the representation (3.3) of $F(t)$ in terms of $\Psi(t)$. As in our calculations of the upper bounds $2\mu_m$ of β (cf. (3.17)), our calculations of the lower bounds ℓ_m of β were *not* carried out to the high accuracy of the calculations of $2nE_{2n}(|x|)$ (95 significant digits) in §1.2. For reasons similar to those applying to the numerical results of §1.3, since our approximations of $\Psi(t)$ are good to about 20 significant digits (cf. [7]), we estimate that the optimization calculations of $\{\ell_m\}_{m=1}^{20}$ are accurate to at least 15 significant digits. The numbers $\{\ell_m\}_{m=1}^{20}$ are given, truncated to 15 significant digits in Table 1.3.

When the upper bounds of Table 1.2 are compared with the lower bounds of Table 1.3, we see that the lower bound ℓ_m of (4.8) is a considerably more *accurate* estimate of β of (1.16) than is the upper bound $2\mu_m$ of (3.17), for *each* $1 \leq m \leq 20$. In fact, the error in ℓ_{20} in approximating β is roughly $9.53 \cdot 10^{-7}$, while that of $2\mu_{100}$ is only $3.88 \cdot 10^{-6}$. Yet, this gain in accuracy was largely offset by the *increased* computer time necessary to find the numbers ℓ_m by our optimization routine. This greater accuracy of the lower bounds ℓ_m in approximating β explains *why* the m values in Table 1.3 do not range as high as those for the upper bounds $2\mu_m$ in Table 1.2.

1.5. The Richardson extrapolation of the numbers $\{2nE_{2n}(|x|)\}_{n=1}^{52}$.

The numbers $\{2nE_{2n}(|x|)\}_{n=1}^{52}$ appearing in Table 1.1 indicate that the convergence of these numbers to the Bernstein constant β (cf. (1.11)) is quite slow. A typical scheme for improving the convergence rate of slowly convergent sequences is the *Richardson extrapolation method* (cf. Brezinski [5, p. 7]), which can be described as follows. If $\{S_n\}_{n=1}^N$, where $N > 2$, is a given (finite) sequence of real numbers, set $T_0^{(n)} := S_n (1 \leq n \leq N)$, and regard $\{T_0^{(n)}\}_{n=1}^N$ as the zeroth column, consisting of N numbers, of the Richardson extrapolation table. The first column of the Richardson extrapolation table, consisting of $N-1$ numbers, is defined by

$$T_1^{(n)} := \frac{x_n T_0^{(n+1)} - x_{n+1} T_0^{(n)}}{x_n - x_{n+1}} \quad (1 \leq n \leq N-1), \tag{5.1}$$

and inductively, the $(k+1)$st column of the Richardson extrapolation table, consisting of $N-k-1$ numbers, is defined by

$$T_{k+1}^{(n)} := \frac{x_n T_k^{(n+1)} - x_{n+k+1} T_k^{(n)}}{x_n - x_{n+k+1}} \quad (1 \leq n \leq N-k-1), \tag{5.2}$$

for each $k = 0, 1, \cdots, N-2$, where $\{x_n\}_{n=1}^{N}$ are given constants. In this way, a triangular table, consisting of $N(N+1)/2$ entries, is created. In our case of $\{2nE_{2n}(|x|)\}_{n=1}^{52}$, a triangular table of 1,378 entries was created. As for the choice of the numbers $\{x_n\}_{n=1}^{52}$ in (5.1)–(5.2), preliminary calculations indicated that $2nE_{2n}(|x|) \doteqdot \beta + K/n^2 +$ lower-order terms, so we chose $x_n := 1/n^2$. We remark that the potential loss of accuracy in the subtractions in the numerators and denominators of the fractions defined in (5.1) and (5.2) *suggested* that the calculations of $2nE_{2n}(|x|)$ be done to very high precision (95 significant digits).

The Richardson extrapolation of $\{2nE_{2n}(|x|)\}_{n=1}^{52}$ produced *unexpectedly beautiful results.* Rather than presenting here the complete extrapolation table of 1,378 entries (giving each entry to, say, 95 significant digits), it seems sufficient to mention that of the last 20 columns of this table, all but 3 of the 210 entries in these columns agreed with the first 45 digits of the following approximation of β in (1.16):

$$\beta \doteq \begin{array}{l} 0.28016 \ 94990 \ 23869 \ 13303 \ 64364 \\ \quad 91230 \ 67200 \ 00424 \ 82139 \ 81236. \end{array} \tag{5.3}$$

In addition, 182 entries of the 210 entries in these 20 last columns of this Richardson extrapolation table were numbers whose first fifty decimal digits *all* agreed with the approximation of β in (5.3)!

1.6. Some open problems.

The success of this Richardson extrapolation (with $x_n := 1/n^2$) applied to $\{2nE_{2n}(|x|)\}_{n=1}^{52}$ strongly suggests that $2nE_{2n}(|x|)$ admits an asymptotic series expansion (cf. Henrici [9, p. 355]) of the form

$$2nE_{2n}(|x|) \overset{?}{\approx} \beta - \frac{K_1}{n^2} + \frac{K_2}{n^4} - \frac{K_3}{n^6} + \cdots \quad (n \to \infty), \tag{6.1}$$

where the constants K_j are independent of n. Assuming that (6.1) is valid, it follows that

$$n^2(2nE_{2n}(|x|) - \beta) \approx -K_1 + \frac{K_2}{n^2} - \frac{K_3}{n^4} + \cdots \quad (n \to \infty). \tag{6.2}$$

Thus, with the known high-precision approximations of $2nE_{2n}(|x|)$ of Table 1.1, and with an estimate for β determined from the last entry of the Richardson extrapolation table for $\{2nE_{2n}(|x|)\}_{n=1}^{52}$, we can again apply the Richardson extrapolation to $\{n^2(2nE_{2n}(|x|) - \beta)\}_{n=1}^{52}$ (with $x_n = 1/n^2$) to obtain an extrapolated estimate for K_1 of (6.2). This bootstrapping procedure can be continued to produce, via Richardson extrapolation, estimates for the successive constants K_j of (6.1). As might be suspected, there is a progressive loss of numerical accuracy in the successive determination of the constants K_j.

In Table 1.4, we tabulate estimates of $\{K_j\}_{j=1}^{10}$, rounded to 10 significant digits. As Table 1.4 indicates, the latter constants K_j begin to grow quite rapidly.

TABLE 1.4
$\{K_j\}_{j=1}^{10}$.

j	K_j
1	0.04396 75288 8
2	0.02640 71687 7
3	0.03125 34264 6
4	0.05889 00165 7
5	0.16010 69971
6	0.59543 53151
7	2.92591 5470
8	18.49414 033
9	146.94301 23
10	1438.03271 7

Because these constants all turned out to be *positive*, we have the following new conjecture.

Conjecture. ([16]). *$2nE_{2n}(|x|)$ admits an asymptotic series expansion of the form*

$$2nE_{2n}(|x|) \overset{?}{\approx} \beta - \frac{K_1}{n^2} + \frac{K_2}{n^4} - \frac{K_3}{n^6} + \cdots \quad (n \to \infty), \tag{6.3}$$

where the constants K_j (independent of n) are all positive.

Next, because the Bernstein constant β has a connection (cf. (3.16)) with particular rational approximations to the function $F(t)$, where $F(t)$ can be expressed (cf. (3.4)) in terms of a classical hypergeometric function, it is not *implausible* to believe that β, as well as the constants K_j of (6.3), might admit a *closed-form expression* in terms of classical hypergeometric functions and/or known mathematical constants.

Finally, Bernstein [2] mentions that his analysis of the behavior of the error $E_{2n}(|x|)$ of best uniform polynomial approximation from π_{2n} to $|x|$ on $[-1,+1]$, extends, for each constant α satisfying $0 < \alpha \leq 1$, to the real-valued even function $f_\alpha(x)$, defined on $[-1,+1]$ by

$$f_\alpha(x) := x^\alpha \text{ for } 0 \leq x \leq 1, \text{ with } f_\alpha(-x) := f_\alpha(x), \tag{6.4}$$

so that $f_1(x) = |x|$. In this case one has, in analogy with (1.11), the *existence* of a constant $\beta(\alpha)$ for each α with $0 < \alpha \leq 1$ for which

$$\lim_{n\to\infty} (2n)^\alpha E_{2n}(f_\alpha) =: \beta(\alpha), \tag{6.5}$$

where

$$E_n(f_\alpha) := \inf\left\{\|f_\alpha(x) - g\|_{L_\infty[-1,+1]} : g \in \pi_n\right\} \qquad (n = 0, 1, \cdots),$$

and $\beta(1)$ is just the Bernstein constant β of (1.11). It is an *open problem* to obtain upper and lower bounds for $\beta(\alpha)$ for, say, $\alpha = \frac{1}{2}$, and to see if

$(2n)^\alpha E_{2n}(f_\alpha)$ possesses an asymptotic series expansion similar to that of (6.3) (for the case $\alpha = 1$).

1.7. Rational approximation of $|x|$ on $[-1, 1]$.

Since the previous sections of this chapter were devoted to the Bernstein Conjecture, i.e., to the problem of best uniform *polynomial* approximation to $|x|$ on $[-1, +1]$, it is natural to finally consider in this chapter the corresponding problem of best uniform *rational* approximation to $|x|$ on $[-1, +1]$. For added notation, let $\pi_{n,n}$ denote, for any nonnegative integer n, that set of all real rational functions $r_{n,n}(x) = p(x)/q(x)$ with $p \in \pi_n$ and $q \in \pi_n$. (Here, it is assumed that p and q have no common factors, that q does not vanish on $[-1, +1]$, and that q is normalized by $q(0) = 1$.) Then for any real-valued function $f(x)$ defined on $[-1, +1]$, we define, in analogy with (1.3),

$$E_{n,n}(f) := \inf \left\{ \|f - r_{n,n}\|_{L_\infty[-1,+1]} : r_{n,n} \in \pi_{n,n} \right\}. \tag{7.1}$$

Interestingly, while Bernstein [2] considered in depth in 1913 the asymptotic behavior of best uniform *polynomial* approximation to $|x|$ on $[-1, +1]$, it was only pointed out fifty years later in 1964 by Newman [12] how decisively *different* best uniform *rational* approximation to $|x|$ on $[-1, +1]$ is, in that Newman constructively showed that

$$\frac{1}{2e^{9\sqrt{n}}} \le E_{n,n}(|x|) \le \frac{3}{e^{\sqrt{n}}} \qquad (n = 4, 5, \cdots). \tag{7.2}$$

Newman's inequalities in (7.2) generated much research interest, and, in the spirit of Bernstein's earlier work on the asymptotic behavior of $E_n(|x|)$ as $n \to \infty$, a good part of this research interest focused on the analogous problem of sharpened asymptotic results for $E_{n,n}(|x|)$, as $n \to \infty$.

For the general theory for the asymptotic behavior of $E_{n,n}(f)$, important contributions have been made by Gonchar [8] and others. For specifically $E_{n,n}(|x|)$, the best results to date have been found by Bulanov [6], who proved that

$$E_{n,n}(|x|) \ge e^{-\pi\sqrt{n+1}} \qquad (n = 0, 1, \cdots), \tag{7.3}$$

and by Vjacheslavov [18], who proved that there exist positive constants M_1 and M_2 such that

$$M_1 \le e^{\pi\sqrt{n}}\, E_{n,n}(|x|) \le M_2 \qquad (n = 1, 2, \cdots). \tag{7.4}$$

Obviously, (7.3) and (7.4) imply both that

$$e^{\pi(1-\sqrt{2})} = 0.27218\cdots \le e^{\pi\sqrt{n}}\, E_{n,n}(|x|) \le M_2 \qquad (n = 1, 2, \cdots), \tag{7.5}$$

and if

$$\underline{M} := \varliminf_{n\to\infty} e^{\pi\sqrt{n}}\, E_{n,n}(|x) \text{ and } \overline{M} := \varlimsup_{n\to\infty} e^{\pi\sqrt{n}}\, E_{n,n}(|x|), \tag{7.6}$$

that

$$1 \leq \underline{M} \leq \overline{M}. \tag{7.7}$$

The result of (7.4) clearly gives the asymptotically *sharp* multiplier, namely π, for $\sqrt{n}$ in the asymptotic behavior of $E_{n,n}(|x|)$ as $n \to \infty$. What only remains then is the determination of the best *asymptotic* constants $\underline{M}$ and $\overline{M}$ in (7.7).

To give insight into this problem, we now describe very recent high-precision calculations of Varga, Ruttan, and Carpenter [17] for the numbers $\{E_{n,n}(|x|)\}_{n=1}^{21}$. As in the polynomial case of §1.2, for each nonnegative integer n, the best uniform approximation to $|x|$ on $[-1,+1]$ from $\pi_{n,n}$, say $\hat{r}_{n,n}(x)$, is unique (cf. [11, p. 158]), so that

$$E_{n,n}(|x|) = \||x| - \hat{r}_{n,n}(x)\|_{L_\infty[-1,+1]} \qquad (n = 1, 2, \cdots). \tag{7.8}$$

Furthermore, since $|x|$ is even in $[-1,+1]$, then so is $\hat{r}_{n,n}(x)$, and this can be shown to imply (cf. (1.5)) that

$$E_{2n,2n}(|x|) = E_{2n+1,2n+1}(|x|) \qquad (n = 1, 2, \cdots). \tag{7.9}$$

Again, it suffices, for our purposes, to consider only the manner in which the sequence $\{E_{2n,2n}(|x|)\}_{n=0}^{\infty}$ decreases to zero.

Next, if $\hat{h}_{n,n}(t) \in \pi_{n,n}$ is the best uniform approximation to $\sqrt{t}$ on $[0,1]$ from $\pi_{n,n}$ for each $n = 1, 2, \cdots$, i.e., if

$$E_{n,n}(\sqrt{t};[0,1]) := \inf_{r_{n,n} \in \pi_{n,n}} \|\sqrt{t} - r_{n,n}(t)\|_{L_\infty[0,1]} = \|\sqrt{t} - \hat{h}_{n,n}(t)\|_{L_\infty[0,1]}, \tag{7.10}$$

then it can be easily shown (cf. (2.4)) that

$$E_{2n,2n}(|x|) = E_{n,n}(\sqrt{t};[0,1]) \qquad (n = 1, 2, \cdots), \tag{7.11}$$

where

$$\hat{r}_{2n,2n}(x) = \hat{h}_{n,n}(x^2) \qquad (n = 1, 2, \cdots). \tag{7.12}$$

From (7.12), our estimates of $\{E_{2n,2n}(|x|)\}_{n=1}^{21}$ were obtained directly from high-precision calculations of $\{E_{n,n}(\sqrt{t};[0,1])\}_{n=1}^{21}$. These calculations, as in §1.2, involved the (second) Remez algorithm, where Brent's MP package [4] was used with up to 150 significant digits, and, allowing for guard digits and the possibility of small rounding errors, we believe that the numbers $\{E_{n,n}(\sqrt{t};[0,1])\}_{n=1}^{21}$ are accurate to 100 significant digits. The numbers $\{E_{2n,2n}(|x|)\}_{n=1}^{21}$ and $\{e^{\pi\sqrt{2n}}E_{2n,2n}(|x|)\}_{n=1}^{20}$ are given in Table 1.5, truncated to 20 digits.

TABLE 1.5

n	$E_{2n,2n}(\|x\|)$	n	$e^{\pi\sqrt{2n}}E_{2n,2n}(\|x\|)$
1	4.36890 12692 07636 1570 (-2)	1	3.71442 65436 83164 1393
2	8.50148 47040 73829 4902 (-3)	2	4.55247 41186 02959 5765
3	2.28210 60097 25259 4879 (-3)	3	5.01604 81727 06945 0372
4	7.36563 61403 07030 5616 (-4)	4	5.32413 85504 99584 3582
5	2.68957 06008 51835 0996 (-4)	5	5.54906 50092 01360 9961
6	1.07471 16229 45128 4948 (-4)	6	5.72308 60623 70144 6149
7	4.60365 92662 63495 9571 (-5)	7	5.86316 39054 52748 1203
8	2.08515 86406 33032 7171 (-5)	8	5.97921 97829 97610 9154
9	9.88933 46452 81424 3884 (-6)	9	6.07751 03145 70501 7015
10	4.87595 75126 31913 2435 (-6)	10	6.16220 95236 00211 8350
11	2.48559 02684 78211 1169 (-6)	11	6.23622 66709 47615 9517
12	1.30437 75913 43073 6526 (-6)	12	6.30166 18824 78634 8671
13	7.02231 99787 39775 6951 (-7)	13	6.36007 54354 31155 6855
14	3.86755 77147 25902 0291 (-7)	14	6.41265 47293 14846 1644
15	2.17398 78201 69794 3205 (-7)	15	6.46032 20136 32057 1274
16	1.24477 08820 89507 1928 (-7)	16	6.50380 62614 76199 8676
17	7.24786 33767 55536 9698 (-8)	17	6.54369 25164 84556 9527
18	4.28546 45582 73508 2156 (-8)	18	6.58045 66245 60485 1075
19	2.56989 67632 18081 6149 (-8)	19	6.61449 02150 91157 3323
20	1.56132 88569 94866 8163 (-8)	20	6.64611 90161 27510 2141
21	9.60112 26128 42236 4808 (-9)	21	6.67561 65126 49122 8856

As in §1.5, we performed several different extrapolation techniques (such as Richardson's extrapolation, Aitken's Δ^2 extrapolation, etc. (cf. Brezinski [5]) on the numbers $\{e^{\pi\sqrt{2n}}\, E_{2n,2n}(|x|)\}_{n=1}^{21}$. Our best results were obtained from Richardson's extrapolation with $x_n = 1/\sqrt{n}$, and, surprisingly, to five *significant digits*, we have obtained that

$$8 \stackrel{?}{=} \lim_{n\to\infty} e^{\pi\sqrt{2n}}\, E_{2n,2n}(|x|). \tag{7.13}$$

The detailed description of the high-precision computations of the numbers $\{E_{2n,2n}(|x|)\}_{n=1}^{21}$ and the extrapolations of $\{e^{\pi\sqrt{2n}}\, E_{2n,2n}(|x|)\}_{n=1}^{21}$ are given in the forthcoming paper of Varga, Ruttan, and Carpenter [17].

REFERENCES

[1] R.A. Bell and S.M. Shah, *Oscillating polynomials and approximations to* $|x|$, Publ. of the Ramanujan Inst., **1**(1969), pp. 167-177.

[2] S. Bernstein, *Sur la meilleure approximation de* $|x|$ *par des polynômes de degrés donnés*, Acta Math. , **37**(1913), pp. 1-57.

[3] R. Bojanic and J.M. Elkins, *Bernstein's constant and best approximation on* $[0, \infty)$, Publ. de l'Inst. Mat., Nouvelle Série, **18**(32)(1975), pp. 19-30.

[4] R. P. Brent, *A* FORTRAN *multiple-precision arithmetic package*, Assoc. Comput. Mach. Trans. Math. Software, **4**(1978), pp. 57-70.

[5] C. Brezinski, **Algorithms d'Accélération de la Convergence**, Éditions Technip, Paris, 1978.

[6] A.P. Bulanov, *Asymptotics for least deviation of* $|x|$ *from rational functions*, (in Russian), Mat. Sbornik, **76**(118)(1968), pp. 288-303. English translation in Math. USSR Sbornik, **5**(1968), pp. 275-290.

[7] W.J. Cody, A.J. Strecok, H.C. Thacher, Jr., *Chebyshev approximations for the psi function*, Math. Comp., **27**(1973), pp. 123-127.

[8] A.A. Gonchar, *Estimates of the growth of rational functions and some of their applications*, (in Russian), Mat. Sbornik, **72**(114)(1967), pp. 489-503. English translation in Math. USSR Sbornik, **1**(1967), pp. 445-456.

[9] P. Henrici, **Applied and Computational Complex Analysis**, Vol. 2, John Wiley & Sons, New York, 1974.

[10] D. Jackson, *Über die Genauigkeit der Annäherung stetiger Funktionen durch rationale Funktionen gegebenen Grades und trigonometrische Summen gegebener Ordnung*, Ph.D. thesis, University of Göttingen, 1911, Göttingen, Germany, 1911.

[11] G. Meinardus, **Approximation of Functions: Theory and Numerical Methods**, Springer-Verlag, Inc., New York, 1967.

[12] D.J. Newman, *Rational approximation to* $|x|$, Michigan Math. J., **11**(1964), pp. 11-14.

[13] E. Ya. Remez, *Sur le calcul effectif des polynômes d'approximation de Tchebichef*, C.R. Acad. Sci. Paris, **199**(1934), pp. 337-340.

[14] T.J. Rivlin, **An Introduction to the Approximation of Functions**, Blaisdell Publishing Co., Waltham, MA, 1969.

[15] D.A. Salvati, *Numerical Computation of Polynomials of Best Uniform Approximation to the Function* $|x|$, Master's thesis, Ohio State University, Columbus, OH, 1980.

[16] R.S. Varga and A.J. Carpenter, *On the Bernstein Conjecture in approximation theory* (in English), Constr. Approx., **1**(1985), pp. 333-348. Russian translation in Mat. Sbornik, **129** (171) (1986), pp. 535-548.

[17] R.S. Varga, A. Ruttan, and A.J. Carpenter, *Numerical results on best uniform rational approximations to* $|x|$ *on* $[-1, +1]$, to appear.

[18] N.S. Vjacheslavov, *On the least deviations of the function sign x and its primitives from the rational functions in the L_p-metrics*, $0 < p \leq \infty$, (in Russian), Mat. Sbornik, **103**(145)(1977), pp. 24-36. English translation in Math. USSR Sbornik, **32**(1977), pp. 19-31.

[19] E.T. Whittaker and G.N. Watson, **A Course of Modern Analysis**, 4th edition, Cambridge University Press, Cambridge, 1962.

CHAPTER 2

The "1/9" Conjecture and Its Recent Solution

2.1. Semi-discrete approximations of parabolic equations.

It is interesting to give here a brief account of the research associated with the so-called "1/9" *Conjecture* and its recent exact solution by Gonchar and Rakhmanov [6]. What is particularly fascinating to us, as we shall see, is that the mathematical research associated with the "1/9" Conjecture, crosscuts the diverse areas of the numerical solution of partial differential equations, matrix theory, rational approximation theory, scientific computation, and complex potential theory.

To begin, consider for simplicity the numerical approximation of the following second-order linear parabolic partial differential equation:

$$\phi u_t(\mathbf{x},t) = \sum_{i=1}^{n} (K_i(\mathbf{x}) u_{x_i}(\mathbf{x},t))_{x_i} - \sigma(\mathbf{x}) u(\mathbf{x},t) + S(\mathbf{x}), \tag{1.1}$$

for all time $t > 0$ and all $\mathbf{x}$ in Ω (where Ω is assumed to be an open, connected, and bounded subset of $\mathbb{R}^n$), subject to the boundary conditions on $\partial\Omega$ (the boundary of Ω) of

$$u(\mathbf{x},t) = g(\mathbf{x}) \qquad (\mathbf{x} \in \partial\Omega;\ t > 0), \tag{1.2}$$

and to the initial conditions of

$$u(\mathbf{x},0) = h(\mathbf{x}) \qquad (\mathbf{x} \in \Omega). \tag{1.3}$$

Here, $S(\mathbf{x})$ and $h(\mathbf{x})$ are given sufficiently smooth functions defined on $\overline{\Omega}$ (the closure of Ω), while $g(\mathbf{x})$ is a given sufficiently smooth function defined on $\partial\Omega$. In addition, ϕ is a given positive constant, and the given quantities $K_i(\mathbf{x})$ and $\sigma(\mathbf{x})$ are assumed to be positive in $\overline{\Omega}$. These assumptions cover important physics and engineering problems. For example, in reactor physics, the time-dependent density of neutrons $u(\mathbf{x},t)$ of a particular average energy in a reactor satisfies (1.1) (in a diffusion theory approximation), where $1/\phi$ physically represents the *average velocity* of these neutrons, and where

$K_i(\mathbf{x}) \equiv K(\mathbf{x})$ and $\sigma(\mathbf{x})$ represent, respectively, the *diffusion coefficient* and the *total removal cross-section* at each point of the reactor. Similar applications extend to petroleum engineering for the flow of a compressible fluid in a homogeneous porous medium.

On using a discrete mesh in $\mathbb{R}^n$ and a suitable $(2n+1)$-point difference approximation to the spatial derivatives, a semi-discrete approximation (i.e., a discrete space but continuous time) of (1.1)–(1.3) gives (cf. Varga [24, p. 253])

$$\text{(1.4)} \qquad \begin{cases} \dfrac{d\mathbf{w}(t)}{dt} &= -A\mathbf{w}(t) + \mathbf{r} \qquad (t > 0), \\ \mathbf{w}(0) &= \tilde{\mathbf{w}}, \end{cases}$$

where A is a sparse real symmetric and positive definite $N \times N$ matrix. The solution of (1.4) can obviously be expressed as

$$\text{(1.5)} \qquad \mathbf{w}(t) = A^{-1}\mathbf{r} + \exp(-tA)\left\{\tilde{\mathbf{w}} - A^{-1}\mathbf{r}\right\} \qquad (t \geq 0),$$

where, as usual, $\exp(-tA) := \sum_{k=0}^{\infty}(-tA)^k/k!$.

It is well known (cf. [24, §8.3]) that the solution in (1.5) is commonly approximated by means of Padé rational approximations of $\exp(-tA)$, and these give, as special cases, the well-known *forward difference, backward difference*, and *Crank–Nicholson* methods for such parabolic problems. Our interest in the next section will be on *Chebyshev*, rather than Padé, rational approximations of $\exp(-tA)$. This is because Padé rational approximations of e^{-x}, being defined as *local* approximations of e^{-x} at $x = 0$, are generally poor approximations of e^{-x} for x large, and this leads to restrictions (for reasons of stability and/or accuracy) on the time steps that can be taken. Chebyshev rational approximations of e^{-x}, in contrast, are defined *globally* with respect to the interval $[0, +\infty)$, and these approximations do not have such time step restrictions, as we shall see.

2.2. Chebyshev semi-discrete approximations.

To define the Chebyshev semi-discrete approximations of (1.5), we consider the following problem in approximation theory. If π_m denotes all real polynomials of degree at most m, and if $\pi_{m,n}$ analogously denotes all real rational functions $r_{m,n}(x) = p(x)/q(x)$ with $p \in \pi_m$ and $q \in \pi_n$, then set

$$\text{(2.1)} \quad \lambda_{m,n} = \lambda_{m,n}(e^{-x}) := \inf\left\{\|e^{-x} - r_{m,n}(x)\|_{L_\infty[0,+\infty)} : r_{m,n} \in \pi_{m,n}\right\}.$$

These constants of (2.1) are called the uniform rational *Chebyshev constants* for e^{-x} on $[0, +\infty)$. It is obvious that $\lambda_{m,n}$ is finite iff $0 \leq m \leq n$. Moreover, given any pair (m, n) of nonnegative integers with $0 \leq m \leq n$, it is known (cf. Achieser [1, p. 55]) that, after dividing out possible common factors, there exists a unique $\hat{r}_{m,n}$ in $\pi_{m,n}$ with

$$\text{(2.2)} \qquad \hat{r}_{m,n}(x) = \hat{p}_{m,n}(x)/\hat{q}_{m,n}(x) \qquad (\hat{p}_{m,n} \in \pi_m; \hat{q}_{m,n} \in \pi_n),$$

and with $\hat{q}_{m,n}(x) > 0$ on $[0, +\infty)$, such that

$$\text{(2.3)} \qquad \lambda_{m,n} = \|e^{-x} - \hat{r}_{m,n}(x)\|_{L_\infty[0,+\infty)}.$$

If $\hat{q}_{m,n}(x) := \sum_{j=0}^{n} c_j x^j$, we set $\hat{q}_{m,n}(tA) := \sum_{j=0}^{n} c_j (tA)^j$, which is a real polynomial of degree n in the $N \times N$ matrix A. Moreover, since $\hat{q}_{m,n}(x) > 0$ on $[0, +\infty)$ and since A is real symmetric and positive definite, we see that the matrix $\hat{q}_{m,n}(tA)$ is also real symmetric and positive definite (and hence nonsingular) for any $t \geq 0$. In analogy with (1.5), we define, (as was originally done in [23]), the (m,n)th *Chebyshev semi-discrete approximation*, $\mathbf{w}_{m,n}(t)$, of the solution $\mathbf{w}(t)$ of (1.5) as

$$(2.4) \quad \mathbf{w}_{m,n}(t) := A^{-1}\mathbf{r} + (\hat{q}_{m,n}(tA))^{-1} (\hat{p}_{m,n}(tA)) \left\{ \tilde{\mathbf{w}} - A^{-1}\mathbf{r} \right\} \quad (t \geq 0),$$

or equivalently, as

(2.4′)

$$\hat{q}_{m,n}(tA)\mathbf{w}_{m,n}(t) = \hat{q}_{m,n}(tA)A^{-1}\mathbf{r} + \hat{p}_{m,n}(tA) \left\{ \tilde{\mathbf{w}} - A^{-1}\mathbf{r} \right\} \quad (t \geq 0).$$

By factoring $\hat{q}_{m,n}(x)$ into its real linear and quadratic factors of x, factors which are also necessarily positive on $[0, +\infty)$, the solution $\mathbf{w}_{m,n}(t)$ of (2.4′) can be achieved by recursively solving a sequence of matrix problems, each problem involving, say, the direct inversion by Gaussian elimination of a linear or quadratic symmetric positive definite matrix in tA, for each $t \geq 0$. (For more details, see Varga [25, p. 71].)

How well does $\mathbf{w}_{m,n}(t)$ of (2.4) approximate $\mathbf{w}(t)$ of (1.5)? On using ℓ_2-vector norms on the N-dimensional complex vector space $\mathbb{C}^N$, it is well known that if B is an Hermitian $N \times N$ matrix with (real) eigenvalues $\{\mu_j\}_{j=1}^{N}$, then the induced ℓ_2-operator norm of B is simply

$$\|B\|_2 = \max_{1 \leq j \leq N} \{|\mu_j|\}.$$

Consequently, if $\{\alpha_j\}_{j=1}^{N}$ similarly denotes the (positive) eigenvalues of the real symmetric positive definite $N \times N$ matrix A in (1.5), then

$$\|\exp(-tA) - \hat{r}_{m,n}(tA)\|_2 = \max_{1 \leq j \leq N} \left| e^{-t\alpha_j} - \hat{r}_{m,n}(t\alpha_j) \right| .$$

But as $t\alpha_j \geq 0$ for all $1 \leq j \leq N$ and for all $t \geq 0$, it follows from (2.1) that

$$\|\exp(-tA) - \hat{r}_{m,n}(tA)\|_2 \leq \lambda_{m,n} \qquad (t \geq 0).$$

Consequently, from (1.5) and (2.4), we have

$$(2.5) \qquad \|\mathbf{w}(t) - \mathbf{w}_{m,n}(t)\|_2 \leq \lambda_{m,n} \|\tilde{\mathbf{w}} - A^{-1}\mathbf{r}\|_2 \qquad (t \geq 0).$$

Note that since the right side of (2.5) is *independent* of t, we have an error bound for $\mathbf{w}(t) - \mathbf{w}_{m,n}(t)$ for *all* $t \geq 0$. In contrast with the familiar Padé methods, which restrict the size of t for reasons of accuracy and/or stability, the Chebyshev semi-discrete method can be used even for very large values of t. In fact, on fixing the integer pair (m,n) with $0 \leq m \leq n$, we can regard the Chebyshev semi-discrete method (2.4) as *defining* a time-stepping procedure with an *arbitrary single positive time step.* (To paraphrase astronaut

Neil Armstrong, "a small step for Padé, a giant leap for Chebyshev!") We do note, however, that Chebyshev semi-discrete methods are *restricted* to linear problems having time-independent coefficients, as in (1.1). For physical problems not of this form, one must use Padé-type approximations of $\exp(-\Delta t A)$. For recent applications of Padé approximations to initial value problems, see Reusch et al. [14].

2.3. "1/9" Conjecture.

The utility of the Chebyshev semi-discrete approximations depends *crucially*, from (2.5), on the behavior of the Chebyshev constants $\lambda_{m,n}$ of (2.1), as $n \to \infty$. From (2.1), it is evident that

$$0 < \lambda_{n,n} \leq \lambda_{n-1,n} \leq \cdots \leq \lambda_{0,n} \qquad (n = 0, 1, \cdots). \tag{3.1}$$

Using an elementary argument, it was shown by Cody, Meinardus, and Varga [5] that if $s_n(x) := \sum_{j=0}^{n} x^j/j!$ denotes the familiar nth partial sum of e^x, then

$$\lim_{n\to\infty} \left\{ \left\| e^{-x} - \frac{1}{s_n(x)} \right\|_{L_\infty[0,+\infty)} \right\}^{1/n} \leq \frac{1}{2}. \tag{3.2}$$

But as $1/s_n(x)$ is an element of $\pi_{0,n}$, (3.2) directly implies from the definition of $\lambda_{0,n}$ in (2.1) that

$$\overline{\lim_{n\to\infty}}\ \lambda_{0,n}^{1/n} \leq \frac{1}{2}. \tag{3.3}$$

Hence, if $\{m(n)\}_{n=0}^{\infty}$ is any sequence of nonnegative integers with $0 \leq m(n) \leq n$, then from (3.3) and the inequalities of (3.1), we have

$$\overline{\lim_{n\to\infty}}\ \left(\lambda_{m(n),n}\right)^{1/n} \leq \frac{1}{2}. \tag{3.4}$$

By means of a shift argument, the upper bound of $\frac{1}{2}$ in (3.4) was slightly improved in the following result.

THEOREM 1. ([5]). *Let* $\{m(n)\}_{n=0}^{\infty}$ *be any sequence of nonnegative integers with* $0 \leq m(n) \leq n$ *for each* $n \geq 0$. *Then,*

$$\frac{1}{6} \leq \overline{\lim_{n\to\infty}}\ \left(\lambda_{m(n),n}\right)^{1/n} \leq \frac{1}{2.29878}. \tag{3.5}$$

The result of (3.5) clearly establishes the *geometric convergence* to zero of the Chebyshev constants $\lambda_{m,n}$ for e^{-x} on $[0, +\infty)$. Since, from (3.1), the sequence $\{\lambda_{n,n}\}_{n=0}^{\infty}$ is necessarily the most *rapidly* converging to zero, Cody, Meinardus, and Varga [5] calculated in 1969 (via the Remez algorithm) the first estimates of $\{\lambda_{n,n}\}_{n=0}^{14}$, which are given in Table 2.1.

TABLE 2.1

n	$\lambda_{n,n}$	$1/\lambda_{n,n}^{1/n}$	n	$\lambda_{n,n}$	$1/\lambda_{n,n}^{1/n}$	n	$\lambda_{n,n}$	$1/\lambda_{n,n}^{1/n}$
0	5.000 (-1)	-	5	9.346 (-6)	10.14	10	1.361 (-10)	9.696
1	6.685 (-2)	14.96	6	1.008 (-6)	9.987	11	1.466 (-11)	9.658
2	7.359 (-3)	11.66	7	1.087 (-7)	9.882	12	1.579 (-12)	9.626
3	7.994 (-4)	10.77	8	1.172 (-8)	9.804	13	1.701 (-13)	9.600
4	8.653 (-5)	10.37	9	1.263 (-9)	9.744	14	1.832 (-14)	9.577

Thus, the rate of geometric convergence to zero of $\lambda_{n,n}$ appeared to be *substantially* better than the upper bound estimate of (3.5). Subsequently, Schönhage [18] proved in 1973 that

$$\frac{1}{6\sqrt{(4n+4)\log 3 + 2\log 2}} \leq 3^n \lambda_{0,n} \leq \sqrt{2} \quad (n = 0, 1, \cdots),$$

so that in fact

$$\lim_{n\to\infty} \lambda_{0,n}^{1/n} = \frac{1}{3}. \tag{3.6}$$

But then, since the number of coefficients available in the rational function $\hat{r}_{n,n}(x)$, which determines $\lambda_{n,n}$, is essentially *twice* the number of coefficients available in $\hat{r}_{0,n}(x)$, which determines $\lambda_{0,n}$, the combination of Schönhage's result (3.6) and the numbers from Table 2.1 (weakly) suggested the following conjecture in 1977:

(3.7) **"1/9" Conjecture.** ([17]). $\lim_{n\to\infty} \lambda_{n,n}^{1/n} \stackrel{?}{=} \frac{1}{9}.$

Next, a numerical *update* of the estimates (from 1969) of $\{\lambda_{n,n}\}_{n=0}^{14}$ of Table 2.1 was carried out in 1984 by Carpenter, Ruttan, and Varga [4], using Richard Brent's MP (multiple precision) package [2] with 230 significant digits. Using the (second) Remez algorithm, as in Chapter 1, these calculations gave the Chebyshev constants $\{\lambda_{n,n}\}_{n=0}^{30}$ to an accuracy of about 200 significant digits. These numbers are given in Table 2.2, rounded to 50 significant digits.

From these numbers, the ratios $\{\lambda_{n-1,n-1}/\lambda_{n,n}\}_{n=1}^{30}$ were computed in [4], and these are given in Table 2.3.

To the last eleven entries of Table 2.3, Richardson's extrapolation was used (with $x_n = 1/n^2$), precisely as in the numerical calculation in Chapter 1 for the Bernstein constant β. These extrapolations are given in Tables 2.4–2.9.

TABLE 2.2
$\{\lambda_{n,n}\}_{n=0}^{30}$.

n	$\lambda_{n,n}$
0	5.000000000000000000000000000000000000000000000(-01)
1	6.683104216185046347061162382711514726145291233 5145(-02)
2	7.358670169580529280012554163080603756744913244 4213(-03)
3	7.993806363356878288081190097111961689765701632 5167(-04)
4	8.652240695288852348224345825414673525007024831 2132(-05)
5	9.345713153026646476753656820792397989608868830 1112(-06)
6	1.008454374899670707934528776410002060407311526 3471(-06)
7	1.087497491375247960866531307272933478485444048 2418(-07)
8	1.172265211633490717795432303938880473510557314 2020(-08)
9	1.263292483322314146094932100909728334334150333 1607(-09)
10	1.361120523345447749870788161536842376472551195 6239(-10)
11	1.466311194937487140668126199557752690348166160 3094(-11)
12	1.579456837051238771486756732818381574685159491 0467(-12)
13	1.701187076340352966416486549945081533337053226 2774(-13)
14	1.832174378254041275155501756513156530559396495 9525(-14)
15	1.973138996612803428625665802082299241769700777 1241(-15)
16	2.124853710495223748799634436418717809044794679 7672(-16)
17	2.288148563247891960405220861269241949471811092 4698(-17)
18	2.463915737765169274831082962323228297774313490 8752(-18)
19	2.653114658063312766926455034695330543463277739 0920(-19)
20	2.856777383549093706690893844930068028829770720 3370(-20)
21	3.076014349505790506914421863975308683947899335 2108(-21)
22	3.312020500551318690751373710822614146028757245 6630(-22)
23	3.566081860636424584769822799765137259723766343 1761(-23)
24	3.839582582168132126936486847301189562943189500 0911(-24)
25	4.134012517285363006270758055452630197056173337 5450(-25)
26	4.450975355730424689793263607279733039511659566 4658(-26)
27	4.792197375888904189931419997885520971051899511 4011(-27)
28	5.159536858257132654665011291255453036410685767 2396(-28)
29	5.554994213751622674642007903879134991027615555 2236(-29)
30	5.980722882849695437271427007124798284642189289 0349(-30)

TABLE 2.3
$\{\lambda_{n-1,n-1}/\lambda_{n,n}\}_{n=1}^{30}$.

n	$\lambda_{n-1,n-1}/\lambda_{n,n}$
1	7.4815532397221509829356536616817047817627984227696(+00)
2	9.0819455991000169708588696090116053013062015321681(+00)
3	9.2054646248528427537813883233351970910096341895810(+00)
4	9.2390013695637342229492228910895668594203903511860(+00)
5	9.2579780201008948071386386966176824253418736773867(+00)
6	9.2673633886078728002406169193047563695298406988185(+00)
7	9.2731650684028757880126091410302398681948193463184(+00)
8	9.2768895688704833336324198589706052242724880476172(+00)
9	9.2794442071765347804120940269531440575883814379160(+00)
10	9.2812683495309755120464682831533111454037182839070(+00)
11	9.2826170054814049413318434721810547537510453765262(+00)
12	9.2836420758101343365763572286526847004054823670187(+00)
13	9.2844394306651793615709775312796518943731046644594(+00)
14	9.2850718606898552364565090058954105627072638256542(+00)
15	9.2855819149043751995853520164465592457350862957931(+00)
16	9.2859992519340952301903430112062879793060351437360(+00)
17	9.2863450591648612312069660869905305380460379908323(+00)
18	9.2866347991400778934888938646138283512490395348046(+00)
19	9.2868799705918004397301050888494834728534387767879(+00)
20	9.2870892682832631479754585814309629455011369059197(+00)
21	9.2872693653333554026168814849193491692724731187991(+00)
22	9.2874254522088778823625744929007529809804239100012(+00)
23	9.2875616152014957847981034264879772236314002103896(+00)
24	9.2876811067903443960632445539185855003199200785395(+00)
25	9.2877865418013514399321174449489929772550616085232(+00)
26	9.2878800417598657237157599122881662160311796009207(+00)
27	9.2879633425048853224353278979631458037287458900494(+00)
28	9.2880378753756707950994314690770008799076337263447(+00)
29	9.2881048291364217038868850048114151489211447279412(+00)
30	9.2881651976905816378400677087169532012794506558118(+00)

TABLE 2.4

First Richardson's extrapolation
9.289026409724499350338080543342629401187948366402 8(+00)
9.2890262501648177327777051096402664917526634197715(+00)
9.2890261238332083354380146232927890779219008634554(+00)
9.2890260227584489781751521371269212102821968923977(+00)
9.2890259411144138331662151027758237265334600018012(+00)
9.2890258745847956720447117179152892402483118606939(+00)
9.2890258199319277513867988095911873373807234645966(+00)
9.2890257746993546055018224389315527078059834110588(+00)
9.2890257370035920984371231103514289542999627148124(+00)
9.2890257053863190014778076067603006933698792208821(+00)

TABLE 2.5

Second Richardson's extrapolation
9.2890254903568100301092506634385383515846398643843(+00)
9.2890254907395748783152475269378170381563748710076(+00)
9.2890254910173236638792234055586598196559281750941(+00)
9.2890254912217618361898849864036303004596910942434(+00)
9.2890254913741950639848522211186105992458585679159(+00)
9.2890254914892118820483798893742287864937739688197(+00)
9.2890254915769523219995629336916174338011290018776(+00)
9.2890254916445664944177855162897305944259530044229(+00)
9.2890254916971628978907097204202614126010390550083(+00)

TABLE 2.6

Third Richardson's extrapolation
9.2890254919264426246904315998037974616346229535605(+00)
9.2890254919246363633882112757200795725544689684432(+00)
9.2890254919235212361782552355452310998144517528836(+00)
9.2890254919227472919000744467255805180069646466419(+00)
9.2890254919222163735816605228071443149565143016341(+00)
9.2890254919218439884743390512914155757325560186146(+00)
9.2890254919215797099009277334795760282585652815175(+00)
9.2890254919213896705910708011872612164000901127673(+00)

TABLE 2.7

Fourth Richardson's extrapolation
9.2890254919205312240649832664389025519177553659037(+00)
9.2890254919208485671587410305609475319973452742869(+00)
9.2890254919207963073654937082427113430339242329911(+00)
9.2890254919208120946294556940428805578880731390886(+00)
9.2890254919208127681771411301709359901892869271757(+00)
9.2890254919208150149547144296258736339472494913060(+00)
9.2890254919208161591023954162336683020414775499319(+00)

TABLE 2.8

Fifth Richardson's extrapolation
9.2890254919214127326587548334445830521388384447460(+00)
9.2890254919206982368598678821050851331068234703891(+00)
9.2890254919208432825305071272154597415183101617465(+00)
9.2890254919208141654584179760993333379827853346193(+00)
9.2890254919208198985165341295732097954739907627740(+00)
9.2890254919208187594380340221604743658919958649908(+00)

TABLE 2.9

Sixth Richardson's extrapolation
9.2890254919196627357020607062507403229154974205964(+00)
9.2890254919210653837136734712907208606483991578876(+00)
9.2890254919207671899154474789653160736121386135206(+00)
9.2890254919208296189900708128670580198741435560234(+00)
9.2890254919208167344095893867600558244128938244875(+00)

The best extrapolated numbers come from Table 2.7, which yields, numerically to 15 significant digits, that

$$\lim_{n\to\infty} \lambda_{n,n}^{1/n} \stackrel{?}{=} \frac{1}{9.28902\ 54919\ 2081}. \tag{3.8}$$

Using a different computational procedure, namely, the Carathéodory–Fejér method, Trefethen and Gutknecht [22] numerically estimated the quantity of (3.8) as follows. Let

$$\exp[(x-1)/(x+1)] = \sum_{k=0}^{\infty}{}' \, c_k T_k(x) \quad (x \in [-1,+1]) \tag{3.9}$$

denote the Chebyshev expansion of $\exp[(x-1)/(x+1)]$ on $[-1,+1]$, where

(3.10)
$$c_k := \frac{2}{\pi}\int_{-1}^{+1} \exp[(x-1)/(x+1)]T_k(x)dx/\sqrt{1-x^2} \quad (k=0,1,\cdots),$$

and where the prime in the summation in (3.9) means that $c_0/2$ is used in place of c_0. From the infinite Hankel matrix $H : [c_{i+j-1}]_{i,j=1}^{\infty}$, let

$$\sigma_n := n\text{th singular value of } H \quad (\text{where } \sigma_1 \geq \sigma_2 \geq \cdots).$$

It was conjectured in [22] that

$$\lambda_{n,n} \overset{?}{\sim} \sigma_n \quad (\text{as } n \to \infty),$$

and, on the basis of numerical estimates of σ_n, Trefethen and Gutknecht [22] conjectured that

(3.11)
$$\lim_{n\to\infty} \lambda_{n,n}^{1/n} \overset{?}{=} \frac{1}{9.28903}.$$

The close numerical estimates of (3.8) and (3.11), based on entirely different numerical methods, gave *strong* evidence that the conjecture of (3.7) is *false.*

There have been a large number of research contributors to the ideas related to the "1/9" Conjecture, and, up to the year 1982, this was surveyed in the monograph of Varga [26]. These research contributions took several distinct directions, the first being to find lower bound estimates of Λ_1 and upper bound estimates of Λ_2, where (cf. (2.1))

(3.12)
$$\Lambda_1 := \varliminf_{n\to\infty} \lambda_{n,n}^{1/n}(e^{-x}) \text{ and } \Lambda_2 := \varlimsup_{n\to\infty} \lambda_{n,n}^{1/n}(e^{-x}),$$

for the geometric convergence rate, by rational approximations, of the *specific* function e^{-x} on $[0,+\infty)$. The best such lower bound for Λ_1 of (3.12) was determined by Schönhage [19] in 1982, who showed that

(3.13)
$$\frac{1}{13.928} < \Lambda_1 := \varliminf_{n\to\infty} \lambda_{n,n}^{1/n}(e^{-x}),$$

and the best upper bound for Λ_2 of (3.9) was determined by Opitz and Scherer [12] in 1985, who showed that

(3.14)
$$\varlimsup_{n\to\infty} \lambda_{n,n}^{1/n}(e^{-x}) =: \Lambda_2 < \frac{1}{9.037}.$$

This result, of course, *proved* that the "1/9" Conjecture of (3.7) is *false.* Actually, (3.14) proves that the *degree* of geometric convergence to zero of the constants $\{\lambda_{n,n}(e^{-x})\}_{n=0}^{\infty}$ is actually *better* than 1/9!

In another direction, it is natural to ask if there are continuous functions $f(x)(\not\equiv 0)$, other than e^x, which are defined on $[0,+\infty)$ for which the Chebyshev constants for $1/f(x)$, namely (cf. (2.1)),

(3.15)
$$\lambda_{m,n}(1/f) := \inf\{\|1/f(x) - r_{m,n}(x)\|_{L_\infty[0,\infty)} : r_{m,n} \in \pi_{m,n}\},$$

possess geometric convergence to zero, i.e., there is a sequence of nonnegative integers $\{m(n)\}_{n=0}^{\infty}$ with $0 \le m(n) \le n$ for all $n \ge 0$, such that

$$\overline{\lim_{n\to\infty}} \left(\lambda_{m(n),n}(1/f)\right)^{1/n} < 1. \tag{3.16}$$

To describe results in this direction, we need the following notation. Given $r > 0$ and $s > 1$, let $\mathcal{E}(r,s)$ denote the closed ellipse in the complex plane, defined by

$$\mathcal{E}(r,s) := \left\{ z = x + iy : \frac{(x-\frac{r}{2})^2}{[\frac{r}{4}(s+\frac{1}{s})]^2} + \frac{y^2}{[\frac{r}{4}(s-\frac{1}{s})]^2} \le 1 \right\}, \tag{3.17}$$

and if $g(z)$ is a function defined on $\mathcal{E}(r,s)$, then set

$$\tilde{M}_g(r,s) := \sup\{|g(z)| : z \in \mathcal{E}(r,s)\}. \tag{3.18}$$

With this notation, we state the result of

THEOREM 2. (Meinardus et al. [9]). *Let $f(x)$ be a real continuous function ($\not\equiv 0$) on $[0,+\infty)$, and assume that there exists a sequence of real polynomials $\{p_n(x)\}_{n=0}^{\infty}$, with $p_n \in \pi_n$ for each $n \ge 0$, and a real number $q > 1$, such that*

$$\overline{\lim_{n\to\infty}} \{\|1/p_n - 1/f\|_{L_\infty[0,\infty)}\}^{1/n} \le \frac{1}{q} < 1. \tag{3.19}$$

Then, $f(x)$ can be extended to an entire function $F(z)$, with $F(x) = f(x)$ for all $x \ge 0$, such that $F(z)$ is of finite order ρ, i.e., if $M_F(r) := \max_{|z|=r}|F(z)|$, then

$$\overline{\lim_{r\to\infty}} \frac{\log\log M_F(r)}{\log r} = \rho, \quad \textit{where} \quad 0 \le \rho < \infty. \tag{3.20}$$

In addition, for every $s > 1$, there exist constants K, θ, and r_0 (all dependent on s and q) such that (cf. (3.18))

$$\tilde{M}_F(r,s) \le K \left(\|f\|_{L_\infty[0,r]}\right)^{\theta} \quad (\textit{for all} \quad r \ge r_0). \tag{3.21}$$

Conversely, let $f(z) = \sum_{k=0}^{\infty} a_k z^k$ be a real entire function with $a_0 > 0$ and $a_k \ge 0$ for all $k \ge 1$. If there exist real numbers $A > 0$, $s > 1$, $\theta > 0$, and $r_0 > 0$ such that (cf. (3.18) *and* (3.21))

$$\tilde{M}_f(r,s) \le A \left(\|f\|_{L_\infty[0,r]}\right)^{\theta} \quad (\textit{for all} \quad r \ge r_0), \tag{3.22}$$

then there exists a sequence of real polynomials $\{p_n(x)\}_{n=0}^{\infty}$, with $p_n \in \pi_n$ for each $n \ge 0$, and a real number $q \ge s^{1/(1+\theta)} > 1$ such that

$$\overline{\lim_{n\to\infty}} \left\{\|1/p_n - 1/f\|_{L_\infty[0,\infty)}\right\}^{1/n} = \frac{1}{q} < 1. \tag{3.23}$$

Consequently, for the choice $m(n) = 0$ for all $n \ge 0$ in (3.15), *then*

$$\overline{\lim_{n\to\infty}} \left(\lambda_{0,n}(1/f)\right)^{1/n} \le \frac{1}{q} < 1. \tag{3.24}$$

It is interesting that the growth rate on ellipses in (3.21) and (3.22) enters in *both* the necessary conditions and sufficient conditions for geometric convergence in Theorem 2. As can be seen, there is a small gap between the necessary conditions and the sufficient conditions of Theorem 2, a gap which is still open.

In yet another related direction taken by researchers in this area was the problem of obtaining geometric convergence to zero, as in (3.2), for uniform rational Chebyshev constants associated with e^{-x}, but now on *domains* different from $[0,+\infty)$, such as *sectors* symmetric in the complex plane about the ray $[0,+\infty)$. To describe such results, for any pair (ν, n) of nonnegative integers, the associated (ν, n)th Padé rational approximation $R_{\nu,n}(x)$ (in $\pi_{\nu,n}$) of e^{-x} is given explicitly (cf. Perron [13, p. 433]) by

$$R_{\nu,n}(x) = Q_{\nu,n}(x)/P_{\nu,n}(x), \tag{3.25}$$

where

$$Q_{\nu,n}(x) := \sum_{j=0}^{\nu} \frac{(n+\nu-j)!\nu!(-z)^j}{(n+\nu)!j!(\nu-j)!}, \text{ and } P_{\nu,n}(z) := \sum_{j=0}^{n} \frac{(n+\nu-j)!n!z^j}{(n+\nu)!j!(n-j)!}. \tag{3.26}$$

Further, for any real number θ with $0 \le \theta < \pi$, define the symmetric sector $S(\theta)$ by

$$S(\theta) := \{z \in \mathbb{C} : |\arg z| \le \theta\}. \tag{3.27}$$

In addition, for any function $h(z)$ defined on $S(\theta)$, we set

$$\|h\|_{L_\infty(S(\theta))} := \sup\{|h(z)| : z \in S(\theta)\}. \tag{3.28}$$

With this notation, we have the result of the following theorem.

THEOREM 3. ([15]). *Assume that the sequence of Padé rational approximants* $\{R_{\nu(n),n}(z)\}_{n=0}^{\infty}$, *as given in* (3.25)–(3.26), *satisfies*

$$\lim_{n\to\infty} \frac{\nu(n)}{n} = \sigma, \quad \textit{where} \quad 0 < \sigma < 1, \tag{3.29}$$

and set

$$\theta_0 := \cos^{-1}\left(\frac{1-\sigma}{1+\sigma}\right), \quad \textit{and} \quad g(\sigma) := \sigma^\sigma(1-\sigma)^{1-\sigma}/2^{1-\sigma}. \tag{3.30}$$

Then, for any θ *satisfying*

$$0 < \theta < 4\tan^{-1}\left\{\left(\frac{1-\sqrt{g(\sigma)}}{1+\sqrt{g(\sigma)}}\right)\tan\left(\frac{\theta_0}{4}\right)\right\}, \tag{3.31}$$

then

(3.32)
$$\overline{\lim_{n\to\infty}} \left\{ \|e^{-z} - R_{\nu(n),n}(z)\|_{L_\infty(S(\theta))} \right\}^{1/n} \leq g(\sigma) \left\{ \frac{\sin[1/4(\theta_0+\theta)]}{\sin[1/4(\theta_0-\theta)]} \right\} < 1.$$

Similar geometric convergence of non-Padé rational approximation to e^{-x} in sectors $S(\theta)$ is given in Saff, Schönhage, and Varga [16].

2.4. Results of Gonchar and Rakhmanov.

In a beautiful and deep new development, Gonchar and Rakhmanov [6] have given an *exact* solution of the "1/9" Conjecture using potential-theoretic methods in the complex plane, methods which unfortunately cannot be adequately described in a few pages. An important role in the development of this theory has been played by results of Nuttal [11] on local rational approximations, based on the theory of Abelian integrals on compact Riemann surfaces, and by the results of Stahl [20] on the asymptotic behavior of multipoint Padé rational approximants. For a survey of these results, see also Stahl [21].

A special case of the results of Gonchar and Rakhmanov [6] is given in the following theorem.

THEOREM 4. (Gonchar and Rakhmanov [6]). *With* $\lambda_{n,n}(e^{-x})$ *defined in* (2.1), *there is a positive number* Λ *with* $0 < \Lambda < 1$ *such that*

$$\lim_{n\to\infty} \left(\lambda_{n,n}(e^{-x})\right)^{1/n} = \Lambda. \tag{4.1}$$

This result of course establishes that the numbers Λ_1 and Λ_2 of (3.12) are *equal.* But what this number Λ numerically is, and how it can be described, is very fascinating!

It turns out that Magnus [8] had earlier correctly identified in 1986 (without a complete proof) that

$$\text{(4.2)}\ \Lambda = \exp(-\pi K'/K) = \frac{1}{9.28902\ 54919\ 20818\ 91875\ 54494\ 35951\cdots},$$

where K and K' are complete elliptic integrals of the first kind for the moduli k and $k' := \sqrt{1-k^2}$, evaluated at the point where $K = 2E, E$ being the complete elliptic integral of the second kind. On the other hand, Gonchar announced, at the International Congress of Mathematicians at Berkeley in August 1986, the following result.

THEOREM 5. (Gonchar and Rakhmanov [6]). *The number* Λ *of* (4.1) *can be characterized in a number-theoretic way as follows. Define*

$$f(z) := \sum_{j=1}^{\infty} a_j z^j, \tag{4.3}$$

where

$$a_j := \left| \sum_{d|j} (-1)^d d \right| \qquad (j = 1, 2, \cdots), \tag{4.4}$$

so that $f(z)$ is analytic in $|z| < 1$. Then, Λ is the unique positive root of the equation

$$f(\Lambda) = \frac{1}{8}. \tag{4.5}$$

Using Newton's method, Carpenter [3] has computed Λ from (4.5) to high precision, and, to 101 significant digits, $1/\Lambda$ is given by

(4.6)

$$\frac{1}{\Lambda} = \begin{array}{l} 9.28902\ 54919\ 20818\ 91875\ 54494\ 35951\ 74506\ 10316\ 94867\ 75012 \\ \quad 44082\ 39700\ 61421\ 72937\ 52472\ 86507\ 07052\ 41587\ 06142\ 47144\cdots, \end{array}$$

which confirms the numerical approximations of (3.8) and (3.11).

In a truly interesting development, Magnus wrote to Gonchar in late 1986 that Λ of (4.1) is also the unique position solution (less than unity) of

$$\sum_{n=0}^{\infty}(2n+1)^2(-\Lambda)^{n(n+1)/2} = 0, \tag{4.7}$$

which is equivalent to the formulation of (4.5), and, moreover, that *exactly one hundred years earlier*, Halphen [7] in 1886 had computed the value of Λ from (4.7) to six significant figures! (Halphen had arrived at the equation in (4.7) in his studies of variations of theta functions.) It is thus fitting and proper that the "1/9" constant be called the *Halphen constant*!

Finally, to connect with the results of Theorems 2 and 3 of §2.3, it is remarked in [6, p. 329] that the main result, Theorem 1, of Gonchar and Rakhmanov [6] also permits the *precise* determination of (i) the degree of best uniform rational approximation of $e^{-p(x)}$ on $[0, +\infty)$, where $p(x)$ is any real polynomial with leading coefficient positive, and (ii) the degree of best uniform rational approximation of e^{-x} in the sector $S(\theta) := \{z : |\arg z| \le \theta\}$ of (3.27), for $\theta < \pi/2$. In the latter case, the result is given in the following theorem.

THEOREM 6. (Gonchar and Rakhmanov [6]). *For each $n = 0, 1, \cdots$, and for $0 \le \theta < \pi/2$, set*

$$\lambda_{n,n}(e^{-x}; S(\theta)) := \inf\{\|e^{-x} - r_{n,n}(x)\|_{L_\infty(S(\theta))} : r_{n,n} \in \pi_{n,n}\}. \tag{4.8}$$

Then, the limit

$$\lim_{n\to\infty}(\lambda_{n,n}(e^{-x}; S(\theta))^{1/n}) = v_\theta \in (0,1) \tag{4.9}$$

exists.

The value of v_θ in (4.9) is explicitly given in [6] in terms of an integral involving theta functions, and, in particular,

$$v_{\pi/4} \doteq \frac{1}{4.42}. \tag{4.10}$$

REFERENCES

[1] N.I. Achieser, **Theory of Approximation**, Frederick Ungar Publishing Co., New York, 1956.

[2] R.P. Brent, *A* FORTRAN *multiple-precision arithmetic package*, Assoc. Comput. Mach. Trans. Math. Software, **4** (1978), pp. 57-70.

[3] A.J. Carpenter, *Some theoretical and computational aspects of approximation theory*, Ph.D. thesis, The University of Leeds, Leeds, England, 1988.

[4] A.J. Carpenter, A. Ruttan, and R.S. Varga, *Extended numerical computations on the* "1/9" *Conjecture in rational approximation theory*, **Rational Approximation and Interpolation**, P.R. Graves-Morris, E.B. Saff, and R.S. Varga, eds., Lecture Notes in Mathematics 1105, Springer-Verlag, Inc., Heidelberg, 1984, pp. 383-411.

[5] W.J. Cody, G. Meinardus, and R.S. Varga, *Chebyshev rational approximation to* e^{-x} *on* $[0, +\infty)$ *and applications to heat-conduction problems*, J. Approx. Theory, **2** (1969), pp. 50-65.

[6] A.A. Gonchar and E.A. Rakhmanov, *Equilibrium distribution and the degree of rational approximation of analytic functions* (in Russian), Mat. Sbornik, **134** (176) (1987), pp. 306-352. English translation in Math. USSR Sbornik, **62** (1989), pp. 305-348.

[7] G.-H. Halphen, **Traité des fonctions elliptiques et de leurs applications**, Gauthier-Villars, Paris, 1886.

[8] A.P. Magnus, CFGT *determination of Varga's constant* "1/9," Preprint B-1348, Inst. Math. Katholieke Univ. Leuven, Louvain, Belgium, 1986.

[9] G. Meinardus, A.R. Reddy, G.D. Taylor, and R.S. Varga, *Converse theorems and extensions in Chebyshev rational approximation to certain entire functions in* $[0, +\infty)$, Trans. Amer. Math. Soc., **170** (1972), pp. 171-186.

[10] G. Meinardus and R.S. Varga, *Chebyshev rational approximations to certain entire functions in* $[0, +\infty)$, J. Approx. Theory, **3** (1970), pp. 300-309.

[11] J. Nuttal, *Asymptotics of diagonal Hermite–Padé polynomials*, J. Approx. Theory, **42** (1984), pp. 299-386.

[12] H.-U. Opitz and K. Scherer, *On the rational approximation of* e^{-x} *on* $[0, \infty)$, Constr. Approx. , **1** (1985), pp. 195-216.

[13] O. Perron, **Die Lehre von den Kettenbrüchen**, Teubner, Leipzig, 1929.

[14] M.F. Reusch, L. Ratzan, N. Pomphrey, and W. Park, *Diagonal Padé approximations for initial value problems*, SIAM J. Sci. Statist. Comput., **9** (1988), pp. 829-838.

[15] E.B. Saff, R.S. Varga, and W.-C. Ni, *Geometric convergence of rational approximations to* e^{-z} *in infinite sectors*, Numer. Math., **26** (1976), pp. 211-225.

[16] E.B. Saff, A. Schönhage, and R.S. Varga, *Geometric convergence to* e^{-x} *by rational functions with real poles*, Numer. Math., **25** (1976), pp. 307-322.

[17] E.B. Saff and R.S. Varga, *Some open questions concerning polynomials and rational functions*, **Padé and Rational Approximation**, E.B. Saff and R.S. Varga, eds., Academic Press, Inc., New York, 1977, pp. 483-488.

[18] A. Schönhage, *Zur rationalen Approximierbarkeit von* e^{-x} *über* $[0, \infty)$, J. Approx. Theory, **7** (1973), pp. 395-398.

[19] ———, *Rational approximations to* e^{-x} *and related* L^2*-problems*, SIAM J. Numer. Anal., **19** (1982), pp. 1067-1082.

[20] H. Stahl, *Orthogonal polynomials with complex-valued weight function*, I, II, Constr. Approx., **2** (1986), pp. 225-240; 241-251.

[21] ———, *General convergence results for rational approximants*, **Approximation VI**, Volume II, C.K. Chui, L.L. Schumaker, and J.D. Ward, eds., Academic Press, Inc., Boston, 1989, pp. 605-634.

[22] L.N. Trefethen and M.H. Gutknecht, *The Carathéodory–Fejér method for real rational approximation*, SIAM J. Numer. Anal., **20** (1983), pp. 420-436.

[23] R.S. Varga, *On higher order stable implicit methods for solving parabolic partial differential equations*, J. Math. and Phys., **XL** (1961), pp. 220-231.

[24] ———, **Matrix Iterative Analysis**, Prentice–Hall, Inc., Englewood Cliffs, NJ, 1962.

[25] ———, **Functional Analysis and Approximation Theory in Numerical Analysis**, CBMS-NSF Regional Conference Series in Applied Math. **3**, Society for Industrial and Applied Mathematics, Philadelphia, PA, 1971.

[26] ———, **Topics in Polynomial and Rational Interpolation and Approximation**, University of Montreal Press, Montreal, 1982.

CHAPTER 3

Theoretical and Computational Aspects of the Riemann Hypothesis

3.1. The Riemann Hypothesis.

Undoubtedly, one of the most famous unsolved problems in mathematics today is the conjecture known as the *Riemann Hypothesis.* For background, the well-known Riemann ζ-function is defined as

$$\zeta(z) := \sum_{n=1}^{\infty} \frac{1}{n^z}, \tag{1.1}$$

and the Dirichlet series of (1.1) converges for any z with $Re\ z > 1$, uniformly, for any fixed $\sigma > 1$, in any subset of $Re\ z \geq \sigma$, which establishes that $\zeta(z)$ is analytic in $Re\ z > 1$. It is also well known that $\zeta(z)$ has the representation

$$\zeta(z) = \prod_{p \text{ a prime}} \left(1 - \frac{1}{p^z}\right)^{-1}, \tag{1.2}$$

which makes $\zeta(z)$ of fundamental interest in analytic number theory. Because the infinite product representation (1.2) is known, for any fixed $\sigma > 1$, to be absolutely convergent in $Re\ z \geq \sigma$, and as no term of this infinite product can vanish in $Re\ z \geq \sigma$, then $\zeta(z)$ is nonzero in $Re\ z > 1$. By means of analytic continuation, it is known (cf. Titchmarsh [32, p. 13]) that $\zeta(z)$ is analytic for all complex numbers z, except for $z = 1$, which is a simple pole of $\zeta(z)$ with residue 1, and that $\zeta(z)$ satisfies the functional equation

$$\zeta(z) = 2^z \pi^{z-1} \sin\left(\frac{z\pi}{2}\right) \Gamma(1-z)\zeta(1-z). \tag{1.3}$$

It follows from (1.3) that $\zeta(z)$ is nonzero in $Re\ z < 0$, except for the simple real zeros $\{-2m\}_{m=1}^{\infty}$, the so-called *trivial zeros* of $\zeta(z)$. It is further known (cf. [32, p. 30]) that $\{-2m\}_{m=1}^{\infty}$ are the only *real* zeros of $\zeta(z)$, and that $\zeta(z)$ possesses infinitely many nonreal zeros which necessarily all lie in the strip $0 \leq Re\ z \leq 1$, the so-called *critical strip* for $\zeta(z)$. In 1859, Riemann [30] postulated his famous conjecture, known as the

(1.4) **Riemann Hypothesis** (1859). *All nonreal zeros of* $\zeta(z)$ *lie on* $Re\ z = \frac{1}{2}$.

Subsequently, it was shown (cf. [32, p. 45]) independently in 1896 by Hadamard and de la Vallée-Poussin that $\zeta(z)$ has no zeros on the line $Re\ z = 1$, which provided the first proof of the famous *prime number theorem*,

$$\pi(x) \sim \frac{x}{\log x} \qquad (x \to +\infty),$$

where $\pi(x)$ denotes the number of primes not exceeding x. Since $\zeta(z) \neq 0$ on $Re\ z = 1$, it follows from (1.3) that $\zeta(x)$ also has no zeros on $Re\ z = 0$; consequently, $\zeta(x)$ possesses infinitely many nonreal zeros in $0 < Re\ z < 1$. Next, it was shown in 1914 by Hardy [12] that $\zeta(z)$ has infinitely many zeros on the *critical line* $Re\ z = \frac{1}{2}$, and in 1974 by Levinson [18] that at least $\frac{1}{3}$ of the zeros of $\zeta(z)$ in the critical strip lie on the critical line. (This fraction has been improved in 1989 to $\frac{2}{5}$ by Conrey [3].)

Throughout the years, even before the advent of supercomputers, numerical calculations were performed to find actual zeros of $\zeta(z)$ in the critical strip. Note that if z is a nonreal zero of $\zeta(z)$, then (1.1) and (1.3) imply that $\bar{z}, 1 - z$, and $1 - \bar{z}$ are also nonreal zeros of $\zeta(z)$. Thus, it suffices to search for nonreal zeros of $\zeta(z)$ only in the *upper* half plane of the critical strip, i.e., in

$$S := \{z \in \mathbb{C} : 0 < Re\ z < 1 \text{ and } Im\ z > 0\}. \tag{1.5}$$

The recent numerical calculations in 1986 by van de Lune, te Riele, and Winter [19] impressively showed that

$$\left\{ \begin{array}{l} \text{in the subset of } S \text{ defined by } 0 < Im\ z \leq 545439823.215, \\ \text{there are precisely } 1,500,000,001 \text{ nonreal zeros of } \zeta(z), \\ \text{which all lie } \textit{exactly} \text{ on } Re\ z = \tfrac{1}{2}, \\ \text{and that each of these zeros is } \textit{simple}. \end{array} \right. \tag{1.6}$$

These computations consumed about 1,500 hours on high-speed computers (primarily the Cyber 205). Equally impressive computations of about 1000 hours on the supercomputer CRAY X-MP were carried out by Odlyzko [22], partially to check the validity of the Riemann Hypothesis but primarily to check some conjectures about the distribution of spacings between successive zeros of $\zeta(z)$ on $Re\ z = \frac{1}{2}$. These computations produced 78,893,234 consecutive zeros of $\zeta(z)$ in the subset of S defined by

$$T := \left\{ \begin{array}{ll} z \in \mathbb{C} : & 0 < Re\ z < 1 \text{ and } \alpha \leq Im\ z \leq \beta, \text{ where} \\ & \alpha := 15202440115916180028.24, \text{ and} \\ & \beta := 15202440115927890387.66. \end{array} \right\} \tag{1.7}$$

All zeros found in this rectangle T were again simple and on the critical line $Re\ z = \frac{1}{2}$. From these extensive computations, *no* counterexample to the

Riemann Hypothesis has to date been found; this, to some, suggests that the Riemann Hypothesis may be true!

There are a number of *different* wide-ranging mathematical techniques which can be employed to attack the truth of the Riemann Hypothesis. In subsequent sections of this chapter, our goal is to focus on the narrower aspects of the connections between the Riemann Hypothesis (formulated in terms of zeros of the Riemann ξ-function), the Pólya Conjecture of 1927 together with its recent solution, and the Laguerre inequalities and Turán inequalities.

3.2. The Pólya Conjecture.

This section is devoted to a conjecture of Pólya from 1927 which is a weaker form of the Riemann Hypothesis. To begin, Riemann's ξ-function can be defined, in the original notation of Riemann [30, p. 147] (as also used in Pólya [25]), by

$$\xi(iz) := \frac{1}{2}\left(z^2 - \frac{1}{4}\right)\pi^{-\frac{z}{2}-\frac{1}{4}}\Gamma\left(\frac{z}{2}+\frac{1}{4}\right)\zeta\left(z+\frac{1}{2}\right), \tag{2.1}$$

where ζ is the Riemann ζ-function of (1.1). (We warn the reader that Titchmarsh [32, p. 16] uses the symbol Ξ for ξ.) Taking into account the known zeros and poles of $\Gamma(z)$ and $\zeta(z)$, it follows that $\xi(z)$ is an entire function, which can be shown to be of order 1 (cf. [32, p. 29]). In addition, Riemann's ξ-function admits the Fourier transform representation (cf. [32, p. 255])

$$\frac{1}{8}\xi\left(\frac{x}{2}\right) = \int_0^\infty \Phi(t)\cos(xt)dt, \tag{2.2}$$

where, if

$$a_n(t) := \left(2\pi^2 n^4 e^{9t} - 3\pi n^2 e^{5t}\right)\exp\left(-\pi n^2 e^{4t}\right) \qquad (n = 1, 2, \cdots), \tag{2.3}$$

then

$$\Phi(t) := \sum_{n=1}^{\infty} a_n(t) \qquad (t \in \mathbb{R}). \tag{2.4}$$

For the reader's convenience and for our subsequent use, we state below some results from Pólya [25] which summarize some known properties of the function $\Phi(t)$. Additional properties of $\Phi(t)$ can be found in Csordas, Norfolk, and Varga [4, Thm. A].

THEOREM 1. (Pólya [25]). *For the function $\Phi(t)$ of* (2.4), *the following are valid:*

(2.5)

$$\begin{cases} (i) & for\ each\ n \geq 1, a_n(t) > 0\ for\ all\ t \geq 0,\ so\ that\ \Phi(t) > 0 \\ & for\ all\ t \geq 0; \\ (ii) & \Phi(z)\ is\ analytic\ in\ the\ strip\ -\pi/8 < Im\ z < \pi/8; \\ (iii) & \Phi(t)\ is\ an\ even\ function,\ so\ that\ \Phi^{(2m+1)}(0) = 0 \quad (m = 0, 1, \cdots); \\ (iv) & for\ any\ \varepsilon > 0,\ \lim_{t\to\infty} \Phi^{(n)}(t)\exp\left[(\pi - \varepsilon)e^{4t}\right] = 0 \quad (n = 0, 1, \cdots). \end{cases}$$

The proofs of statements (2.5i)–(2.5iv) can be found in Pólya [25]. Except for (2.5iii), which is by no means obvious, the above properties of $\Phi(t)$ can be established in a straightforward way.

Returning to the integral representation of (2.2), on expanding $\cos(xt)$ and integrating termwise, the Maclaurin series for $\xi\left(\frac{x}{2}\right)/8$ is seen to be

$$\frac{1}{8}\xi\left(\frac{x}{2}\right) = \sum_{m=0}^{\infty} \frac{\hat{b}_m(-x^2)^m}{(2m)!}, \tag{2.6}$$

where the moments $\hat{b}_m$ of $\Phi(t)$ are defined by

$$\hat{b}_m := \int_0^{\infty} t^{2m}\Phi(t)dt \qquad (m = 0, 1, \cdots). \tag{2.7}$$

On setting $z = -x^2$ in (2.6), the function $F_0(z)$ is then defined by

$$F_0(z) := \sum_{m=0}^{\infty} \frac{\hat{b}_m z^m}{(2m)!}, \tag{2.8}$$

so that from (2.6),

$$\frac{1}{8}\,\xi\,\left(\frac{x}{2}\right) = F_0(-x^2). \tag{2.9}$$

Because ξ is an entire function of order 1, (2.9) shows that F_0 is an *entire function of order* $\frac{1}{2}$. Moreover, as the moments $\hat{b}_m$ are all *positive* from (2.5i), F_0 is a *real* entire function (i.e., $F_0(z)$ is real for real z), with $F_0(x) > 0$ for all $x \geq 0$.

To connect the function F_0 of (2.8) with the Riemann Hypothesis of (1.4), it is interesting to first mention that the *original* formulation of the Riemann Hypothesis (cf. Riemann [30, p. 148]) was that all zeros of ξ (cf. (2.1)) *are real.* From (2.9), we see that x_0 is a zero of $\xi\left(\frac{x}{2}\right)$ iff $z_0 := -x_0^2$ is a negative real zero of $F_0(z)$. Thus, it follows that *the truth of the Riemann Hypothesis is equivalent to the statement that all zeros of* F_0 *are real and negative.*

Next, it is convenient to introduce the following class of real entire functions. With the usual convention that $\prod_{j=1}^{0} := 1$, consider any real entire function of the form

$$f(z) = Ce^{-\lambda z^2+\beta z} z^n \prod_{j=1}^{\omega}\left(1 - \frac{z}{x_j}\right) e^{z/x_j} \qquad (0 \leq \omega \leq \infty), \tag{2.10}$$

where $\lambda \geq 0, \beta$ and $C \neq 0$ are real constants, n is a nonnegative integer, and the x_j's are real and nonzero with $\sum_{j=1}^{\omega} x_j^{-2} < \infty$. The collection of all such functions which can be represented in this form is called the *Laguerre–Pólya class*, and we write $f \in \mathcal{L} - \mathcal{P}$ if f has the form (2.10). The following result, essentially due to Laguerre [17, p. 174] (cf. Boas [1, p. 24], gives, in (2.11), the classical *Laguerre inequalities*.

THEOREM 2. (Laguerre [17]). *Let f be an element of $\mathcal{L}-\mathcal{P}$. Then,*
(2.11)

$$L_{p+1}(x;f) := (f^{(p+1)}(x))^2 - f^{(p)}(x)f^{(p+2)}(x) \geq 0 \quad (x \in \mathbb{R}, p = 0, 1, \cdots).$$

Proof. The logarithmic derivative of f is, from (2.10),

$$\frac{f'(z)}{f(z)} = -2\lambda z + \beta + \frac{n}{z} + \sum_{j=1}^{\omega} \left\{ \frac{1}{(z-x_j)} + \frac{1}{x_j} \right\},$$

from which it follows, using the hypothesis that $f \in \mathcal{L}-\mathcal{P}$, that
(2.12)

$$\left(\frac{f'(x)}{f(x)}\right)' = \frac{f(x)f''(x) - (f'(x))^2}{f^2(x)} = -\left[2\lambda + \frac{n}{x^2} + \sum_{j=1}^{\omega} \frac{1}{(x-x_j)^2}\right] \leq 0,$$

i.e.,

$$(f'(x))^2 - f(x)f''(x) \geq 0 \quad (x \in \mathbb{R}). \tag{2.13}$$

It is further known (cf. Pólya and Schur [27]) that f in $\mathcal{L}-\mathcal{P}$ implies that f' is also in $\mathcal{L}-\mathcal{P}$, which inductively gives (2.11). □

If the Maclaurin expansion of f in $\mathcal{L}-\mathcal{P}$ is $f(z) = \sum_{m=0}^{\infty} c_m z^m$, then applying (2.11) only at $x = 0$ clearly gives

$$mc_m^2 \geq (m+1)c_{m-1}c_{m+1} \quad (m = 1, 2, \cdots). \tag{2.14}$$

But if $\lambda > 0$ or if $\omega = \infty$ for the given f in $\mathcal{L}-\mathcal{P}$, (2.12) gives strict inequality in (2.13) and hence in (2.14), i.e.,

$$mc_m^2 > (m+1)c_{m-1}c_{m+1} \quad (m = 1, 2, \cdots). \tag{2.15}$$

This can be used as follows.

Since F_0 is a real entire function of order $\frac{1}{2}$, the truth of the Riemann Hypothesis would imply that the infinitely many zeros of F_0 are all real and negative. As such, the stronger result (2.15), applied to F_0 of (2.8), is valid, and this gives

$$\left(\hat{b}_m\right)^2 > \left(\frac{2m-1}{2m+1}\right)\hat{b}_{m-1}\hat{b}_{m+1} \quad (m = 1, 2, \cdots), \tag{2.16}$$

or equivalently, that

$$D_m := \left(\hat{b}_m\right)^2 - \left(\frac{2m-1}{2m+1}\right)\hat{b}_{m-1}\hat{b}_{m+1} > 0 \quad (m = 1, 2, \cdots). \tag{2.17}$$

Thus, we see that *all* the inequalities of (2.16) are *necessary* conditions for the truth of the Riemann Hypothesis. In contemporary terminology, the inequalities (2.16) are commonly called *Turán inequalities*, while the numbers D_m of (2.17) are called *Turán differences*.

In 1925-1926, shortly after the death of Jensen in 1925, Pólya was granted permission to study the unpublished notes of Jensen. This was of great importance to the mathematical world since Jensen had announced 14 years earlier that he would publish his new results on necessary and sufficient conditions for the truth of the Riemann Hypothesis, but these results were never published. Pólya materially extended these unpublished notes, and from them published a paper [25] in 1927, which included these previously unpublished necessary and sufficient conditions of Jensen for the truth of the Riemann Hypothesis. In this paper, Pólya used the notation

$$b_m := \frac{2 \cdot m!}{(2m)!} \hat{b}_m \qquad (m = 0, 1, \cdots),$$

where $\hat{b}_m$ is defined in (2.7), so that, in Pólya's notation, the weaker form (2.14) applied to F_0, gives

$$(\dagger) \qquad b_n^2 - b_{n-1} b_{n+1} \geq 0 \qquad (n = 1, 2, \cdots).$$

In [25, p. 16], Pólya also remarked:*

> How little, however, we have progressed in this direction is most clearly shown [by the fact] that the next-to-simplest necessary condition (†) for the reality of the zeros of the ξ-function is not yet verified.

This remark has become known as

The Pólya Conjecture (1927).

$$(2.18) \qquad \left(\hat{b}_m\right)^2 \overset{?}{>} \left(\tfrac{2m-1}{2m+1}\right) \hat{b}_{m-1} \hat{b}_{m+1} \quad (m = 1, 2, \cdots).$$

The interest in the Pólya Conjecture is very natural: if one of the inequalities (2.18) were to *fail* for some $m \geq 1$, or equivalently, if D_m of (2.17) were *nonpositive* for some $m \geq 1$, then the Riemann Hypothesis *would be false.*

The history concerning the Pólya Conjecture of 1927 is worth recounting. For nearly 40 years, this problem apparently remained untouched in the literature. Then in 1966 and 1969, Grosswald [10], [11] generalized a formula of Hayman [13] on admissible functions, and, as an application of this generalization, Grosswald proved, in the notation of (2.16) and (2.17), that

$$(2.19) \qquad \begin{aligned} D_m :=& \left(\hat{b}_m\right)^2 - \left(\tfrac{2m-1}{2m+1}\right) \hat{b}_{m-1} \hat{b}_{m+1} \\ =& \frac{\left(\hat{b}_m\right)^2}{m} \left\{1 + O\left(\tfrac{1}{\log m}\right)\right\} \quad (m \to \infty). \end{aligned}$$

---

* "Wie wenig wir aber an diesem Wege fortgeschritten sind, zeigt es am deutlichsten, dass die nächsteinfachste notwendige Bedingung für die Realität aller Nullstellen

$$(17) \qquad b_n^2 - b_{n-1} b_{n+1} \geq 0$$

im Falle der ξ-Funktion noch nicht verifiziert ist."

As the moments $\hat{b}_m$ are necessarily positive from (2.5*i*) for all $m \geq 0$, Grosswald's result (2.19) proves that (2.16) *is* valid for all m sufficiently large, say, $m \geq m_0$. Unfortunately, the value of m_0 was not determined from his analysis, and, to our knowledge, this gap in Grosswald's solution of the Pólya Conjecture was not subsequently filled in the literature.

The delicate nature of the Turán inequalities (2.16) can be seen from the following calculation. As $\Phi(t)$ is positive for all $t \geq 0$ from (2.5*i*), we can write the moment $\hat{b}_m$ of (2.7) in the form

$$\hat{b}_m^2 = \left\{ \int_0^\infty t^{(2m-2)/2} \sqrt{\Phi(t)} \cdot t^{(2m+2)/2} \sqrt{\Phi(t)} dt \right\}^2 .$$

Applying the Cauchy–Schwarz inequality to this integral directly gives the inequality $(\hat{b}_m)^2 \leq \hat{b}_{m-1}\hat{b}_{m+1}$, which we equivalently write as

$$\hat{b}_m^2 \leq \left(\frac{2m+1}{2m+1} \right) \hat{b}_{m-1}\hat{b}_{m+1} \qquad (m = 1, 2, \cdots), \tag{2.20}$$

whereas the sought Turán inequalities (2.16) are nearly the reversed inequalities:

$$\hat{b}_m^2 > \left(\frac{2m-1}{2m+1} \right) \hat{b}_{m-1}\hat{b}_{m+1} \qquad (m = 1, 2, \cdots).$$

The Pólya Conjecture fascinated us (i.e., Csordas, Norfolk, and Varga), largely because the Turán inequalities (2.11) had not been checked numerically in the literature. Certainly, high-precision calculations of the moments $\hat{b}_m$ of (2.7) did not seem formidable, because of the extremely rapid decay of $\Phi(t)$ to zero as $t \to \infty$ from (2.5*iv*). Thus, in 1983 we enthusiastically began our high-precision calculations (to 50 significant digits) of the moments $\{\hat{b}_m\}_{m=0}^{109}$, from which the Turán differences $\{D_m\}_{m=1}^{108}$ were then determined. There was, after all, a *remote chance* of uncovering a $D_m \leq 0$.

But, alas, our calculations produced no surprising results: *all computed* D_m *were positive.* We list in Table 3.1 the numbers $\{\hat{b}_m\}_{m=0}^{20}$, $\{D_m\}_{m=1}^{20}$ and $\{\tilde{D}_m\}_{m=1}^{20}$, where

$$\tilde{D}_m := \frac{mD_m}{(\hat{b}_m)^2} \qquad (m = 1, 2, \cdots), \tag{2.21}$$

so that (cf. (2.19))

$$\tilde{D}_m = 1 + O\left(\frac{1}{\log m} \right) \qquad (m \to \infty). \tag{2.21$'$}$$

All numbers in Table 3.1 have been truncated to 11 significant digits. Since the moments $\hat{b}_m$ are all positive, we see from (2.21) that $\tilde{D}_m$ is positive iff the Turán difference D_m is positive. Note that the $\tilde{D}_m$'s in Table 3.1 are much more slowly varying, as a function of m, than the corresponding D_m's, which is consistent with the result of (2.21$'$).

It might appear from Table 3.1 that the moments $\hat{b}_m$ are *strictly decreasing* for $m \geq 0$. It is easily shown that this cannot be the case, and it became an intriguing little mathematical puzzle to determine, *both* numerically as well as mathematically (using asymptotic methods), the exact value of m for which $\hat{b}_m$ is minimized. It turns out (cf. Csordas, Norfolk, and Varga [5]) that $\hat{b}_m$ is strictly decreasing for $0 \leq m \leq 339$, and is strictly increasing for $m \geq 339$, so that

$$\min_{m \geq 0} \hat{b}_m = \hat{b}_{339} = 2.18540\ 10467 \cdots 10^{-71}. \tag{2.22}$$

TABLE 3.1

m	$\hat{b}_m$	D_m	$\tilde{D}_m$
0	6.21400 97273 (-2)	— —	— —
1	7.17873 25984 (-4)	3.58844 91486 (-8)	6.96323 80609 (-2)
2	2.31472 53388 (-5)	3.16329 93950 (-11)	1.18078 64542 (-1)
3	1.17049 98956 (-6)	7.05673 24419 (-14)	1.54519 91985 (-1)
4	7.85969 60229 (-8)	2.83222 02230 (-16)	1.83389 94114 (-1)
5	6.47444 26609 (-9)	1.73636 66894 (-18)	2.07112 67219 (-1)
6	6.24850 92806 (-10)	1.47803 17201 (-20)	2.27134 00906 (-1)
7	6.85711 35660 (-11)	1.64153 36845 (-22)	2.44379 71530 (-1)
8	8.37956 28564 (-12)	2.27744 38477 (-24)	2.59474 65240 (-1)
9	1.12289 59005 (-12)	3.82273 77260 (-26)	2.72858 83427 (-1)
10	1.63076 65724 (-13)	7.57537 75877 (-28)	2.84852 92500 (-1)
11	2.54307 50583 (-14)	1.73849 34268 (-29)	2.95697 31508 (-1)
12	4.22669 38654 (-15)	4.54925 56467 (-31)	3.05576 53485 (-1)
13	7.44135 71845 (-16)	1.34019 54348 (-32)	3.14635 11038 (-1)
14	1.38066 04233 (-16)	4.39776 86757 (-34)	3.22988 20677 (-1)
15	2.68793 65964 (-17)	1.59301 19382 (-35)	3.30728 97809 (-1)
16	5.47056 43869 (-18)	6.32085 57309 (-37)	3.37933 76956 (-1)
17	1.16018 31858 (-18)	2.72899 35268 (-38)	3.44665 87993 (-1)
18	2.55669 85949 (-19)	1.27457 93250 (-39)	3.50978 33473 (-1)
19	5.84001 96623 (-20)	6.40679 74312 (-41)	3.56915 96545 (-1)
20	1.37967 28720 (-20)	3.45025 04583 (-42)	3.62516 99207 (-1)

Finally, two comments should be made in this section. First, the details of just how the moments $\{\hat{b}_m\}_{m=0}^{109}$ were computed will be given in §3.6. Second, numerically determining the numbers of Table 3.1 was *not a wasted effort*. As we shall see in §3.3, these numerical calculations gave the motivation for a *rigorous* mathematical proof of the truth of the Pólya Conjecture, and more!

3.3. The resolution of the Pólya Conjecture.

In 1986, Csordas, Norfolk, and Varga [4] gave a lengthy construction (consisting of 12 lemmas) of the fact that if (cf. (2.4))

$$\Psi(t) := \int_t^{\infty} \Phi\left(\sqrt{u}\,\right) du \qquad (t \geq 0), \tag{3.1}$$

then

$$\log \Psi(t) \quad \textit{is strictly concave on} \quad (0, +\infty). \tag{3.2}$$

With (3.2), the *truth of the Pólya Conjecture* was then established in [4], which settled in a constructive manner the nearly 60-year-old Pólya Conjecture!

For our purposes here, we describe below a subsequent result of Csordas and Varga [7] which is similar to (3.2), but is the basis for a stronger result than the truth of the Pólya Conjecture. We begin with the following theorem.

THEOREM 3. ([7]). *The function* $\Phi(t)$ *of* (2.4) *has the property that*

$$\log \Phi\left(\sqrt{t}\right) \quad \textit{is strictly concave on} \quad (0, +\infty). \tag{3.3}$$

Proof. (Sketch). From (2.5*ii*) of Theorem 1, $\Phi(z)$ is analytic in the strip $-\pi/8 < Im\ z < \pi/8$, which implies that $\Phi(t) \in C^{\infty}(\mathbb{R})$, and since $\Phi(t) > 0$ for all $t \geq 0$ from (2.5*i*), it is readily verified that $\left(d^2/dt^2\right) \log \Phi\left(\sqrt{t}\ \right) < 0$ on $(0, +\infty)$ iff $g(t) > 0$ on $(0, +\infty)$, where

$$g(t) := t\left[\left(\Phi'(t)\right)^2 - \Phi(t)\Phi''(t)\right] + \Phi(t)\Phi'(t) \qquad (t \geq 0). \tag{3.4}$$

In [7], a lengthy construction (consisting of 10 lemmas which we do not reproduce here) established bounds for $\Phi^{(j)}(t)$ $(j = 1, 2, \cdots, 6)$ on different intervals which resulted in

$$\begin{cases} g(t) > 0 & (0 < t \leq 0.03), \\ g(t) > 0 & (0.03 \leq t \leq 0.06), \\ g(t) > 0 & (0.056 \leq t < \infty), \end{cases} \tag{3.5}$$

which gives the desired result of (3.3). $\square$

We mention at this point that earlier in 1982, unknown to us, Matiyasevich [20] had expressed the Turán difference D_m of (2.17) as the following useful triple integral:

$$2(2m+1)D_m = \int_0^\infty \int_0^\infty u^{2m} v^{2m} \Phi(u)\Phi(v) \left[\left(v^2 - u^2\right) \int_u^v -\frac{d}{dt}\left(\frac{\Phi'(t)}{t\Phi(t)}\right) dt\right] du\ dv, \tag{3.6}$$

for any $m \geq 1$, which, with the definition of $g(t)$ in (3.4), can be also expressed as

$$2(2m+1)D_m = \int_0^\infty \int_0^\infty u^{2m} v^{2m} \Phi(u)\Phi(v) \left[\left(v^2 - u^2\right) \int_u^v \frac{g(t)dt}{t^2\Phi^2(t)}\right] du\ dv, \tag{3.7}$$

for any $m \geq 1$. Since $g(t)$ and $\Phi(t)$ are both positive on $(0, +\infty)$ from (3.5) and (2.5*i*), it is clear that the quantity in brackets in (3.7) is nonnegative for all $0 \leq u, v < \infty$. But as the remaining terms in the integrand of (3.7) are also nonnegative, then $D_m > 0$ for all $m \geq 1$. Thus, as a consequence (cf. (2.12) and (2.13)), *the Pólya Conjecture is true.*

Interestingly, Matiyasevich [20] claimed that $g(t)$ of (3.4) is positive on $(0, +\infty)$ "from interval computations that are as powerful as a proof," and, while interval arithmetic computations, showing that $g(t) > 0$, similarly break down the interval $(0, +\infty)$ into parts, as is done in (3.5), we did not find enough information in [20] to completely verify his computations. However, his claim that $g(t) > 0$ on $(0, +\infty)$ is of course *true*, since there is a constructive proof of this in [7], as well as for the logarithmic concavity of $\Psi(t)$ of (3.1) in [4]. We emphasize that our approach in resolving the Pólya Conjecture was based on the *independent* notion that $\log \Phi\left(\sqrt{t}\,\right)$ is strictly concave on $(0, +\infty)$, and this will be used in proving the more general result, Theorem 4, of this section.

Next, assume that $K(t) : \mathbb{R} \to \mathbb{R}$ satisfies

$$(3.8) \qquad \begin{cases} (i) & K \text{ is integrable over } \mathbb{R}, \\ (ii) & K(t) > 0 \qquad (t \in \mathbb{R}), \\ (iii) & K(t) = K(-t) \qquad (t \in \mathbb{R}), \text{ and} \\ (iv) & \text{for some } \varepsilon > 0, K(t) = O(\exp(-|t|^{2+\varepsilon})) \qquad (t \to \infty). \end{cases}$$

Such a function $K(t)$ is called an *admissible kernel.* It is known (cf. Pólya [26]) that the Fourier transform

$$(3.9) \qquad H(x, K) := \int_{-\infty}^{+\infty} K(t) e^{ixt} dt = 2 \int_0^{\infty} K(t) \cos(xt) dt$$

of an admissible kernel $K(t)$ is a real entire function of finite order ρ, where, with the ε of (3.8 *iv*), this order ρ satisfies

$$(3.10) \qquad \rho \leq \frac{2+\varepsilon}{1+\varepsilon} < 2.$$

Moreover, if $K(t)$ is an admissible kernel and if $H(x, K)$ of (3.9) has only *real* zeros, then another beautiful result of Pólya [26] gives that the entire function

$$(3.11) \qquad J(x; K, f) := \int_{-\infty}^{+\infty} f(it) K(t) e^{ixt} dt \qquad (f \in \mathcal{L} - \mathcal{P})$$

also has only real zeros. That is, in Pólya's terminology (cf. [26, p. 7]), the functions $f(it)$ are *universal factors* which preserve the reality of the zeros of the entire function $H(x, K)$ of (3.9). (In fact, Pólya showed in [26] that the functions $f(it)$, where $f \in \mathcal{L} - \mathcal{P}$, are the only *analytic* functions which satisfy this property!)

The above can be applied as follows. First, it is evident from (2.5) that $\Phi(t)$ satisfies (3.8), so that $\Phi(t)$ is an admissible kernel. Next, we have, in the notation of (3.9), that (cf. (2.2))

$$(3.12) \qquad H(x, \Phi) = 2 \int_0^{\infty} \Phi(t) \cos(xt) dt = \frac{1}{4} \xi\left(\frac{x}{2}\right).$$

Thus, the truth of the Riemann Hypothesis would imply that $H(x, \Phi)$ has only real zeros. Now, consider any *even* f in $\mathcal{L}-\mathcal{P}$, i.e., from (2.10),

$$f(z) = Ce^{-\lambda z^2} z^{2n} \prod_{j=1}^{\omega} \left(1 - \frac{z^2}{x_j^2}\right) \qquad (0 \leq \omega \leq \infty), \tag{3.13}$$

where $\lambda \geq 0$, C is a nonzero real constant, n is a nonnegative integer, and the x_j's satisfy $x_j > 0$ with $\sum_{j=1}^{\omega} 1/x_j^2 < \infty$. But, up to a sign change, it can be easily verified that $f(it)\Phi(t)$ is an admissible kernel for *any* even f in $\mathcal{L}-\mathcal{P}$ (cf. (3.13)), the proof making use of the properties of (2.5). Consequently,

$$J(x; \Phi, f) := 2 \int_0^{\infty} f(it)\ \Phi(t) \cos(xt) dt, \tag{3.14}$$

is a real entire function of order less than 2 (from (3.10)). Moreover, from our discussion above, the truth of the Riemann Hypothesis would imply that $J(x; \Phi, f)$ has only real zeros for any even f in $\mathcal{L}-\mathcal{P}$. Now, the Maclaurin expansion of $J(x; \Phi, f)$ is just

$$J(x; \Phi, f) = \sum_{m=0}^{\infty} \frac{\hat{b}_m(f)(-x^2)^m}{(2m)!}, \tag{3.15}$$

where

$$\hat{b}_m(f) := 2 \int_0^{\infty} t^{2m} f(it)\Phi(t) dt \qquad (m = 0, 1, \cdots), \tag{3.16}$$

and, on setting $z = -x^2$ in (3.15), we obtain, in analogy with (2.8), the real entire function

$$F(z; f) := \sum_{m=0}^{\infty} \frac{\hat{b}_m(f) z^m}{(2m)!}, \tag{3.17}$$

whose order is less than unity from (3.10). But then, on applying Theorem 2 to $F(z; f)$, the truth of the Riemann Hypothesis would imply that

(3.18)

$$D_m(f) := \left(\hat{b}_m(f)\right)^2 - \left(\frac{2m-1}{2m+1}\right) \hat{b}_{m-1}(f)\hat{b}_{m+1}(f) > 0 \quad (m = 1, 2, \cdots),$$

for any such *even* $f(z)$ in $\mathcal{L}-\mathcal{P}$. We note that (3.18) is a *generalized* form of the Pólya Conjecture of (2.18), since the choice $f(z) \equiv 1$ in (3.18) reduces to (2.18).

Based on the logarithmic concavity of $\Phi\left(\sqrt{t}\,\right)$ in (3.3) of Theorem 3, we now establish the following result of Csordas and Varga [7].

THEOREM 4. ([7]). *For any real even entire function of the form*

$$h(z) := Ce^{-\lambda z^2} z^{2n} \prod_{j=1}^{\omega} \left(1 - \frac{z^2}{x_j^2}\right) \qquad (0 \leq \omega \leq \infty), \tag{3.19}$$

where λ *and* $C \neq 0$ *are real constants,* n *is a nonnegative integer, and the* x_j*'s are real and nonzero with* $\sum_{j=1}^{\omega} 1/x_j^2 < \infty$, *set*

$$\hat{b}_m(h) := 2\int_0^\infty t^{2m} h(it)\Phi(t)dt \qquad (m = 0, 1, \cdots). \tag{3.20}$$

Then, the associated Turán differences, defined by

$$D_m(h) := \left(\hat{b}_m(h)\right)^2 - \left(\frac{2m-1}{2m+1}\right)\hat{b}_{m-1}(h)\hat{b}_{m+1}(h) \quad (m = 1, 2, \cdots), \tag{3.21}$$

satisfy

$$D_m(h) > 0 \qquad (m = 1, 2, \cdots). \tag{3.22}$$

Proof. By hypothesis, the kernel

$$K(t) := h(it)\Phi(t) \qquad (t \in \mathbb{R}) \tag{3.23}$$

is (up to a sign change) an admissible kernel, and it suffices from (3.7) and (3.3) to show that

$$\log K\left(\sqrt{t}\right) \text{ is strictly concave on } (0, +\infty). \tag{3.24}$$

Clearly,

$$K\left(\sqrt{t}\right) = \tilde{C}e^{\lambda t}t^n \prod_{j=1}^{\omega}\left(1 + \frac{t}{x_j^2}\right)\Phi\left(\sqrt{t}\right) \qquad \left(\tilde{C} := (-1)^n C\right),$$

so that

$$\log K\left(\sqrt{t}\right) = \log\tilde{C} + \lambda t + n\log t + \sum_{j=1}^{\omega}\log\left(1 + \frac{t}{x_j^2}\right) + \log\Phi\left(\sqrt{t}\right).$$

On differentiating, we obtain

$$\frac{d^2}{dt^2}\log K\left(\sqrt{t}\right) = -\left[\frac{n}{t^2} + \sum_{j=1}^{\omega}\frac{1}{(x_j^2+t)^2}\right] + \frac{d^2}{dt^2}\log\Phi\left(\sqrt{t}\right) \qquad (t > 0).$$

As the term in brackets above is nonnegative for all $t > 0$, and as the final term above is negative from (3.3) for all $t > 0$, then (3.24) is established. □

It is important to note that Theorem 4 allows *all* real λ, whereas the preceding discussion on even universal factors allows only $\lambda \geq 0$.

The results of Theorem 4 are also applicable in fairly general situations. Indeed, consider any entire function of the form

$$\hat{H}(x) := \int_0^\infty K(t)\cos(xt)dt, \tag{3.25}$$

where $K(t)$ is any $C^2(\mathbb{R})$ function which is an admissible kernel (cf. (3.8)). Let the moments corresponding to the function $K(t)$ be defined as

$$\hat{c}_m := \int_0^\infty t^{2m} K(t) dt \qquad (m = 0, 1, 2, \cdots). \tag{3.26}$$

Then, a *necessary condition* for the entire function $\hat{H}(x)$ to have only real zeros is that

$$\hat{c}_m^2 > \left(\frac{2m-1}{2m+1}\right) \hat{c}_{m-1}\hat{c}_{m+1} \qquad (m = 1, 2, \cdots). \tag{3.27}$$

By Theorem 4, a *sufficient condition* for (3.27) to hold is that

$$\frac{d^2}{dt^2} \log\left(K\left(\sqrt{t}\,\right)\right) < 0 \qquad (t > 0). \tag{3.28}$$

As an example of how (3.28) can be applied, consider, as in [7], the function $\hat{K}(t) := \exp(-2\cosh(t))$. Then it is known (cf. Pólya [24]) that the cosine transform of $\hat{K}(t)$, namely,

$$\int_0^\infty \exp(-2\cosh(t)) \cos(xt) dt,$$

is a real entire function having only real zeros. Since

$$\log \hat{K}\left(\sqrt{t}\,\right) = -2\cosh\left(\sqrt{t}\,\right) \qquad (t \geq 0),$$

then

$$\frac{d^2}{dt^2} \log \hat{K}\left(\sqrt{t}\,\right) = \frac{1}{2}\left\{\frac{\sinh\left(\sqrt{t}\,\right)}{t^{3/2}} - \frac{\cosh\left(\sqrt{t}\,\right)}{t}\right\} \qquad (t > 0).$$

But, as the Maclaurin expansion (in the variable $u := \sqrt{t}$) of the quantity in braces has all *negative* coefficients, we see that the sufficient condition (3.28) for the inequalities of (3.27) to hold, is satisfied. It is interesting to remark that the kernel $\hat{K}(t) = \exp(-2\cosh(t))$ *cannot* be expressed as $f(it)$ where $f(z)$ is of the form (2.10), so that this example does not involve even universal factors.

As another application of the previous results, we have the following corollary.

COROLLARY 5. ([7]). *Let*

$$K_\lambda(t) := \Phi(t)\cosh(\lambda t) \qquad (\lambda \in \mathbb{R}), \tag{3.29}$$

where $\Phi(t)$ *is defined in* (2.4), *and let*

$$\hat{c}_m(\lambda) := \int_0^\infty t^{2m} K_\lambda(t) dt \qquad (m = 0, 1, \cdots). \tag{3.30}$$

Then,

$$(3.31)\quad (\hat{c}_m(\lambda))^2 > \left(\frac{2m-1}{2m+1}\right)\hat{c}_{m-1}(\lambda)\hat{c}_{m+1}(\lambda) \quad (m = 1, 2, \cdots; \lambda \in \mathbb{R}).$$

Proof. To deduce the inequalities of (3.31), it is sufficient to have

$$\frac{d^2}{dt^2}\log K_\lambda\left(\sqrt{t}\,\right) = \frac{d^2}{dt^2}\log\Phi\left(\sqrt{t}\,\right) + \frac{d^2}{dt^2}\log\left(\cosh\left(\lambda\sqrt{t}\,\right)\right) < 0 \quad (0 < t < \infty).$$

But with (3.3), it is clearly sufficient to have $\frac{d^2}{dt^2}\log\left(\cosh\left(\lambda\sqrt{t}\,\right)\right) \le 0$ for all $0 < t < \infty$ and all $\lambda \in \mathbb{R}$. As an easy calculation shows, this is true iff (with $u := \sqrt{t}$)

$$(3.32)\quad \sigma_\lambda(u) := -\lambda^2 u + \lambda\sinh(2\lambda u)/2 \ge 0 \qquad (0 < u < \infty; \lambda \in \mathbb{R}).$$

But as $\sigma_\lambda(0) = 0$ and $\sigma'_\lambda(u) = \lambda^2\{-1 + \cosh(2\lambda u)\} \ge 0$, then (3.32) is valid. □

The inequalities (3.31) are known to be true in the special cases $\lambda \ge 1$ (cf. Pólya [24, p. 32]) and $\lambda = 0$ (cf. [4]). For $\lambda = 1$, the kernel $K_1(t)$ of (3.29) is of particular interest since Pólya showed in [25] that the Fourier cosine transform of $K_1(t)$, i.e.,

$$(3.33)\qquad F_1(x) := \int_0^\infty K_1(t)\cos(xt)dt,$$

has only real zeros. Pólya's method of proof also shows that the entire function

$$F_\lambda(x) := \int_0^\infty K_\lambda(t)\cos(xt)dt$$

has only real zeros if $\lambda \ge 1$, and, consequently, (3.31) holds for all $\lambda \ge 1$.

3.4. The de Bruijn–Newman constant Λ.

We recall that the Riemann ξ-function admits the Fourier transform representation (cf. (2.2))

$$(4.1)\qquad \frac{1}{8}\xi\left(\frac{x}{2}\right) = \frac{1}{2}\int_{-\infty}^{\infty}\Phi(t)e^{ixt}dt = \int_0^\infty \Phi(t)\cos(xt)dt,$$

where

$$(4.2)\qquad \Phi(t) := \sum_{n=1}^{\infty}\left(2\pi^2 n^4 e^{9t} - 3\pi n^2 e^{5t}\right)\exp\left(-\pi n^2 e^{4t}\right).$$

For any real λ, we introduce the factor $e^{\lambda t^2}$ in the integrand in (4.1) and we set

$$(4.3)\qquad H_\lambda(x) := \frac{1}{2}\int_{-\infty}^{+\infty} e^{\lambda t^2}\Phi(t)e^{ixt}dt = \int_0^\infty e^{\lambda t^2}\Phi(t)\cos(xt)dt,$$

so that from (2.2), $H_0(x) = \xi(x/2)/8$. On expanding $\cos(xt)$, the Maclaurin expansion for $H_\lambda(x)$ is given by

$$H_\lambda(x) = \sum_{m=0}^{\infty} \frac{\hat{b}_m(\lambda)(-x^2)^m}{(2m)!}, \tag{4.4}$$

where the moments $\hat{b}_m(\lambda)$ are defined by

$$\hat{b}_m(\lambda) := \int_0^\infty t^{2m} e^{\lambda t^2} \Phi(t) dt \qquad (m = 0, 1, \cdots; \lambda \in \mathbb{R}). \tag{4.5}$$

It is known from Csordas, Norfolk, and Varga [6, App. A] that $H_\lambda(x)$ is a real entire function of order 1 (and type ∞) for *each* real λ. Moreover, since $\Phi(t) > 0$ for all real t from (2.5i), then $\hat{b}_m(\lambda) > 0$ for all $m = 0, 1, \cdots$ and all real λ. Consequently, $H_\lambda(x)$ of (4.4) is a *real* entire function of order 1, for all real λ.

Our aim in this section is to consider the *behavior* of the zeros of $H_\lambda(x)$, as a function of the real parameter λ, and to relate this to the Riemann Hypothesis. To begin, we note that $g(t) := e^{-\lambda t^2}$, for $\lambda \geq 0$, is (cf. (2.10)) an element of $\mathcal{L} - \mathcal{P}$, so that

$$g(it) = e^{\lambda t^2} \qquad (\lambda \geq 0)$$

is a *universal factor* in Pólya's terminology. Hence from Pólya [26], if $H_0(x) = \xi(x/2)/8$ has only real zeros, then so does $H_\lambda(x)$ of (4.3) for any $\lambda \geq 0$. Subsequently, two results of de Bruijn [2] establish that

$$\text{(4.6)} \quad \begin{array}{ll} (i) & H_\lambda(x) \textit{ has only real zeros for any } \lambda \geq \tfrac{1}{2}; \\ (ii) & \textit{if } H_\lambda(x) \textit{ has only real zeros for some real } \lambda, \\ & \textit{then } H_{\lambda'}(x) \textit{ has only real zeros for any } \lambda' \geq \lambda. \end{array}$$

Specifically, the truth of the Riemann Hypothesis would imply that $H_\lambda(x)$ has only real zeros for any $\lambda \geq 0$.

It is interesting that Newman [21] has shown more recently that a real number Λ exists which satisfies $-\infty < \Lambda \leq \frac{1}{2}$, such that

$$\text{(4.7)} \quad \begin{cases} H_\lambda(x) \textit{ has only real zeros when } \lambda \geq \Lambda, \textit{ and} \\ H_\lambda(x) \textit{ has some nonreal zeros when } \lambda < \Lambda. \end{cases}$$

Because of de Bruijn's earlier related work, it is fitting to call this constant Λ the *de Bruijn–Newman constant.*

Newman's method of proof of the *existence* of Λ in [21] was nonconstructive, so that no information, such as an explicit *lower bound* for Λ, appears in [21]. As for *upper bounds* for Λ, certainly $\Lambda \leq \frac{1}{2}$ is valid, and the truth of the Riemann Hypothesis would imply $\Lambda \leq 0$. (Newman [21] offers the complementary conjecture that $\Lambda \geq 0$, with the statement "that this new conjecture is a quantitative version of the dictum that the Riemann Hypothesis, if true, is only barely so.") Thus, we were led to the challenging new problem (related to the Riemann Hypothesis) of numerically determining upper and lower bounds for the de Bruijn–Newman constant Λ. At the time of this writing, only numerical results giving *lower bounds* for Λ have recently been considered, and we report on this activity below. Before discussing these newly found lower bounds for Λ, we provide some useful theoretical results for $\mathcal{L}-\mathcal{P}$ functions.

First, setting $z = -x^2$ in (4.4) serves to define the function $F_\lambda(z)$:

$$F_\lambda(z) := \sum_{m=0}^{\infty} \frac{\hat{b}_m(\lambda) z^m}{(2m)!}, \tag{4.8}$$

where $\hat{b}_m(\lambda)$ is given in (4.5). Since $H_\lambda(x)$ of (4.4) and $F_\lambda(z)$ of (4.8) are related through

$$H_\lambda(x) = F_\lambda(-x^2) \qquad (\lambda \in \mathbb{R}), \tag{4.9}$$

then $F_\lambda(z)$ is evidently a real entire function of order $\frac{1}{2}$ which necessarily has (cf. Boas [1, p. 24]) infinitely many (with possibly some nonreal) zeros, for each choice of real λ. In addition, since the $\hat{b}_m(\lambda) > 0$ imply F_λ is positive on the ray $z \geq 0$, then from (4.7) and (4.9),

$$\begin{cases} F_\lambda(z) \textit{ has only real negative zeros when } \lambda \geq \Lambda, \textit{ and} \\ F_\lambda(z) \textit{ has some nonreal zeros when } \lambda < \Lambda. \end{cases} \tag{4.10}$$

But, from the definition of the Laguerre–Pólya class of functions in (2.10), (4.10) can be succinctly expressed as

$$\begin{cases} F_\lambda \in \mathcal{L}-\mathcal{P} \textit{ when } \lambda \geq \Lambda, \textit{ and} \\ F_\lambda \notin \mathcal{L}-\mathcal{P} \textit{ when } \lambda < \Lambda, \end{cases} \tag{4.11}$$

which can be used as follows.

Consider an arbitrary element $G(z)$ in $\mathcal{L}-\mathcal{P}$, which we write in the form

$$G(z) = \sum_{m=0}^{\infty} \frac{\gamma_m}{m!} z^m, \tag{4.12}$$

and let $G_n(z)$ denote its associated nth *Jensen polynomial*:

$$G_n(t) := \sum_{k=0}^{n} \binom{n}{k} \gamma_k\, t^k \qquad (n = 0, 1, \cdots). \tag{4.13}$$

A known result of Pólya and Schur [27] states that

$$G(z) \in \mathcal{L}-\mathcal{P} \text{ iff } G_n(t) \in \mathcal{L}-\mathcal{P} \qquad (n = 0, 1, \cdots), \tag{4.14}$$

or equivalently,

$$G(z) \in \mathcal{L}-\mathcal{P} \text{ iff } G_n(t) (\not\equiv 0) \text{ has only real zeros }, \tag{4.15}$$

for all $n = 1, 2, \cdots$.

With (4.15), we immediately have the result of Proposition 6.

PROPOSITION 6. ([6]). *Suppose that there is a positive integer m and a real number λ such that the mth Jensen polynomial for $F_\lambda(z)$ of* (4.8), *given by*

$$G_m(t;\lambda) := \sum_{k=0}^{m} \binom{m}{k} \frac{\hat{b}_k(\lambda)\cdot k!}{(2k)!} t^k, \tag{4.16}$$

has a nonreal zero. Then,

$$\lambda < \Lambda. \tag{4.17}$$

Proposition 6 then provides us with the basis for the following numerical algorithm to find lower bounds for the de Bruijn–Newman constant Λ:

(*i*) fix a real $\lambda < 0$;

(*ii*) with (4.5), determine high-precision estimates $\{\beta_k(\lambda)\}_{k=0}^{N}$ of the moments $\left\{\hat{b}_k(\lambda)\right\}_{k=0}^{N}$;

(*iii*) with (*ii*), and (4.16), form the *approximate Jensen polynomials*

$$g_m(t;\lambda) := \sum_{k=0}^{m} \binom{m}{k} \frac{\hat{\beta}_k(\lambda)k!}{(2k)!} t^k \qquad (m = 1, 2, \cdots, N); \tag{4.18}$$

(*iv*) with a high-precision polynomial root-finder, find the zeros of $g_m(t;\lambda)$ $(m = 1, 2, \cdots, N)$;

(*v*) if there is a positive integer m (with $1 \le m \le N$) for which the approximate mth Jensen polynomial $g_m(t;\lambda)$ of (*iii*) has a nonreal zero, then prove theoretically that the corresponding exact Jensen polynomial $G_m(t;\lambda)$ is *guaranteed* to also have a nonreal zero. Then, $\lambda < \Lambda$.

The word "guaranteed," as used in the previous paragraph, can surely strike fear into an analyst's heart! We discuss this more carefully in the next section.

3.5. Lower bounds for Λ via the Jensen polynomials.

In Csordas, Norfolk, and Varga [6], we implemented Proposition 6 to find a particular lower bound for the de Bruijn–Newman constant Λ. Specifically, on choosing $\lambda = -50$, we numerically computed the moments $\{\hat{b}_m(-50)\}_{m=0}^{16}$, via the *Romberg integration method* (cf. Stoer and Bulirsch [31, p. 136]), to a relative accuracy of at least 60 significant digits, and the 16th associated approximate Jensen polynomial $g_{16}(t;-50)$ of (4.18), had the nonreal zero

$$z_1 := -220.91911\ 17368\ 44951\cdots + i7.09256\ 52553\ 63889\ 67\cdots, \tag{5.1}$$

with modulus

$$|z_1| = 221.03293\ 51307\ 13450\cdots . \tag{5.2}$$

In order to show that the (exact) Jensen polynomial G_{16} $(t; -50)$ of (4.16) is also *guaranteed* to have a nonreal zero, the following result of Ostrowski [23, App. B] was used in [6].

PROPOSITION 7. (Ostrowski [23]). *Let* $f(z) = \sum_{j=0}^{n} a_j z^j$ *(with* $a_0 a_n \neq 0$*) and* $h(z) = \sum_{j=0}^{n} b_j z^j$ *be two complex polynomials, and let the zeros of* $f(z)$ *be* $\{z_j\}_{j=1}^{n}$ *(which are all nonzero, but multiple zeros are allowed). Assume that there is a positive real number* τ *with* $4n\tau^{1/n} \leq 1$ *such that*

$$|b_j - a_j| \leq \tau a_j \qquad (j = 0, 1, \cdots, n). \tag{5.3}$$

Then, the n zeros $\{w_j\}_{j=1}^{n}$ *of* $h(z)$ *can be ordered in such a way that*

$$|w_j - z_j| < 8n\tau^{1/n}|z_j| \qquad (j = 1, 2, \cdots, n). \tag{5.4}$$

Using Proposition 7, the following result was obtained in [6].

THEOREM 8. ([6]). *If* Λ *is the de Bruijn–Newman constant, then*

$$-50 < \Lambda. \tag{5.5}$$

Proof. Set $f(z) := g_{16}(z; -50) = \sum_{k=0}^{16} \binom{16}{k} \frac{\hat{\beta}_k(-50)k!z^k}{(2k)!}$ and $h(z) := G_{16}(z; -50) = \sum_{k=0}^{16} \binom{16}{k} \frac{\hat{b}_k(-50)k!z^k}{(2k)!}$. Then, as the accuracy of the approximate (numerically calculated) moments $\hat{\beta}_k(-50)$, in relationship to the exact moments $b_k(-50)$ implies that

$$|\hat{b}_k(-50) - \hat{\beta}_k(-50)| \leq 10^{-60}\hat{\beta}_k(-50) \qquad (k = 0, 1, \cdots, 16),$$

we see that (5.3) is satisfied with $\hat{\tau} := 10^{-60}$, so that $8n\hat{\tau}^{1/n} \leq 2.27620 \cdot 10^{-2}$ for $n = 16$. Then, (5.4) of Proposition 7 gives for z_1 of (5.1) that there is a zero, w_1, of $G_{16}(z; -50)$ such that

$$|w_1 - z_1| \leq \left(2.27620 \cdot 10^{-2}\right)|z_1| < 5.03116. \tag{5.6}$$

But as $|Im\ w_1 - Im\ z_1| \leq |w_1 - z_1|$, the above inequality, with (5.1), gives

$$|Im\ w_1 - 7.09256\cdots| < 5.03116,$$

so that

$$Im\ w_1 > 2.06164. \tag{5.7}$$

Thus, $G_{16}(z; -50)$ possesses a nonreal zero, and applying Proposition 6 gives the desired result. □

The lower bound of (5.5) for Λ was the *first* such lower bound determined. As mentioned in [6], the importance of this result was not the specific lower bound, but rather that a *constructive method* had been found to obtain lower bounds for Λ. It was also stated in this paper that there was optimism for finding improved lower bounds with this method.

It turns out that Ostrowski's Proposition 7, which gives a *global* relative accuracy between *all* the zeros of two nearby polynomials, is far too *conservative* for the purpose, as used in the proof of Theorem 8, of comparing two particular *simple* (i.e., not multiple) zeros of two nearby polynomials. In this regard, we give the following elementary, but useful, result.

LEMMA 9. *Let $p(z)$ be a complex polynomial of degree n. If $p'(z_1) \neq 0$, then the disk*

$$|z - z_1| \leq n|p(z_1)|/|p'(z_1)| \tag{5.8}$$

contains at least one zero of $p(z)$.

Proof. As the result of Lemma 9 is trivial if $p(z_1) = 0$, assume $p(z_1) \neq 0$, and write $p(z) = \gamma \prod_{k=1}^{n} (z - \zeta_k)$, where the ζ_k's are the zeros of $p(z)$. Taking the logarithmic derivative of $p(z)$ and evaluating the result in the point z_1 gives

$$\frac{p'(z_1)}{p(z_1)} = \sum_{k=1}^{n} \frac{1}{z_1 - \zeta_k}.$$

On taking absolute values in the above expression, we have

$$\frac{|p'(z_1)|}{|p(z_1)|} \leq \sum_{k=1}^{n} \frac{1}{|z_1 - \zeta_k|} \leq \frac{n}{\min\limits_{1 \leq k \leq n} |z_1 - \zeta_k|},$$

which, on rewriting, is just (5.8). □

We remark that the result of Lemma 9 is a special case of a more general result (cf. Henrici [14, p. 454]).

Without going into details, using Lemma 9, rather than Proposition 7, requires far *fewer* digits of accuracy in calculating the moments $\hat{b}_m(\lambda)$ to guarantee that $G_{16}(z; -50)$ has a nonreal zero. With this, in Varga, Norfolk, and Ruttan [33], the successively improved lower bounds for Λ of Table 3.2 were determined, where the third column of this table is the degree of the first Jensen polynomial for which a nonreal zero can be *guaranteed.*

TABLE 3.2

λ	degree n	digits required	zero
-100	10	12	$-453.840\cdots + i9.703\cdots$
$-\ 50$	16	12	$-220.919\cdots + i7.092\cdots$
$-\ 20$	41	18	$-111.065\cdots + i1.322\cdots$
$-\ 15$	56	20	$-79.834\cdots + i0.282\cdots$
$-\ 10$	97	21	$-45.530\cdots + i.0.156\cdots$
$-\ 8$	142	21	$-30.993\cdots + i0.124\cdots$

As λ increases to zero, Table 3.2 indicates a rapid increase in the *degree* of the Jensen polynomial, which gives rise to a nonreal zero, as well as a marked increase in the *significant digits needed* (column 3). To underscore this, te Riele [29], using the Jensen polynomials approach but with a modification involving Sturm sequences, recently reported strong numerical evidence for the lower bound

$$-5 < \Lambda, \tag{5.9}$$

based on a Jensen polynomial of degree 406 and 250 digits of accuracy used in the associated computations!

In the next section, a different procedure is used, whose implementation gives a much improved lower bound for the de Bruijn–Newman constant Λ.

3.6. Tracking zeros of $F_\lambda(z)$.

It is evident from the definition in (4.8) that if, for some real $\lambda < 0, F_\lambda(z)$ of (4.8) possesses a nonreal zero, then $F_\lambda(z) \notin \mathcal{L}-\mathcal{P}$ from (4.11); whence, $\lambda < \Lambda$. Since $F_\lambda(z)$ is an entire function of order $\frac{1}{2}$, it necessarily possesses infinitely many zeros (cf. Boas [1, p. 24]). The idea now is to *directly track* particular zeros of $F_\lambda(z)$. We note for further use that $F_\lambda(z)$ of (4.8) can be expressed, in analogy with (4.3), in integral equation form as

$$F_\lambda(z) = \int_0^\infty e^{\lambda t^2}\Phi(t)\cosh(t\sqrt{z})dt \qquad (\lambda \in \mathbb{R}). \tag{6.1}$$

Now, suppose, for λ_0 real, that $z(\lambda_0)$ is some *simple* zero of $F_{\lambda_0}(z)$, so that $z(\lambda)$ remains a simple zero of $F_\lambda(z)$ in some small interval in λ containing λ_0 in its interior. In this interval, $F_\lambda(z(\lambda)) \equiv 0$, so that, with the definition of the moments $\hat{b}_m(\lambda)$ of (4.5),

$$\begin{aligned} F_\lambda(z(\lambda)) \equiv 0 \quad &= \int_0^\infty e^{\lambda t^2}\Phi(t)\cosh\left(t\sqrt{z(\lambda)}\,\right)dt \\ &= \sum_{m=0}^\infty \frac{\hat{b}_m(\lambda)}{(2m)!}(z(\lambda))^m . \end{aligned} \tag{6.2}$$

On differentiating (6.2) with respect to λ, we obtain

$$0 \equiv \sum_{m=0}^\infty \frac{\hat{b}_{m+1}(\lambda)(z(\lambda))^m}{(2m)!} + \frac{dz(\lambda)}{d\lambda}\sum_{m=0}^\infty \frac{(m+1)\hat{b}_{m+1}(\lambda)(z(\lambda))^m}{(2m+2)!}.$$

Because the sum above which multiplies $dz(\lambda)/d\lambda$ is nonzero as $z(\lambda)$ is assumed to be a simple zero, then solving for $dz(\lambda)/d\lambda$ yields

$$\frac{dz(\lambda)}{d\lambda} = -\frac{\displaystyle\sum_{m=0}^\infty \hat{b}_{m+1}(\lambda)(z(\lambda))^m/(2m)!}{\displaystyle\sum_{m=0}^\infty (m+1)\hat{b}_{m+1}(\lambda)(z(\lambda))^m/(2m+2)!}. \tag{6.3}$$

It is also important to note that since $\hat{b}_m(\lambda) := \int_0^\infty t^{2m}e^{\lambda t^2}\Phi(t)dt$, then replacing $e^{\lambda t^2}$ by its Maclaurin expansion and integrating termwise, gives

$$\hat{b}_m(\lambda) = \sum_{j=0}^\infty \frac{\hat{b}_{m+j}(0)\lambda^j}{j!} \qquad (m = 0, 1, \cdots; \lambda \in \mathbb{R}). \tag{6.4}$$

Thus, for λ small and negative, one needs from (6.4) to compute *only one extended table* of high-precision moments $\{\hat{b}_m(0)\}_{m=0}^{N}$, from which the moments $\{\hat{b}_m(\lambda)\}_{m=0}^{N'}$ can be directly estimated, using (6.4). In fact, using the Encore parallel computer in the Department of Mathematics and Computer Sciences at Kent State University, the moments $\{\hat{b}_m(0)\}_{m=0}^{1600}$ were each computed to an accuracy of 220 significant digits, using simply the trapezoidal rule with a sufficiently fine mesh. More precisely, if $T_h(m)$ denotes the trapezoidal rule approximation (on a uniform mesh of size h) of $\hat{b}_m(0)$, then from (2.5*ii*), it can be shown (cf. [33]) that

$$\left|T_h(m) - \hat{b}_m(0)\right| = O\left(\exp\left(\frac{-(\pi^2 - \varepsilon)}{4h}\right)\right) \qquad (h \to 0).$$

It is this geometric convergence which makes this application of the trapezoidal rule both fast and accurate.

Next, let $\{\rho_j := \frac{1}{2} + it_j\}_{j=1}^{15,000}$ denote the known simple zeros of the ζ-function on the critical line $Re\ z = \frac{1}{2}$, which are tabulated in te Riele [28]. By virtue of our change of variables $z = -x^2$, it follows that $\{z_j(0) := -4t_j^2\}_{j=1}^{15,000}$ are then zeros of $F_0(z)$. It turns out that *certain* pairs of successive zeros of $F_0(z)$ are quite close. One such pair of simple zeros, which we focused on, is

$$\text{(6.5)} \qquad z_{34}(0) = -49,310.231\cdots, \text{ and } z_{35}(0) = -50,063.757\cdots,$$

and, with the formula of (6.3), it was determined that

$$\text{(6.6)} \qquad \left.\frac{dz_{34}(\lambda)}{d\lambda}\right|_{\lambda=0} = +877.836\cdots, \text{ and } \left.\frac{dz_{35}(\lambda)}{d\lambda}\right|_{\lambda=0} = -26.627\cdots.$$

Because of the *difference* in signs of these two derivatives in (6.6), this means that these two zeros are *tending toward one another* as λ decreases from zero, i.e., these two zeros are *attracted* to each other.

This tracking of pairs of zeros of $F_\lambda(z)$ generates some interesting geometrical results. In Figure 3.1, we have graphed the 21 pairs of zeros

$$\{z_{34}(-[0.04]j) \text{ and } z_{35}(-[0.04]j)\}_{j=0}^{20}.$$

We see that the zeros $z_{34}(\lambda)$, and $z_{35}(\lambda)$ start out as real zeros which move toward one another. These zeros then *meet*, forming a *double zero* of $F_\lambda(z)$ when $\lambda \doteqdot -0.38$, and then these zeros bifurcate into a nonreal conjugate complex pair of zeros which follow a parabolic-like trajectory in the complex plane. We see from Figure 3.1 that this pair of zeros, $z_{34}(\lambda)$ and $z_{35}(\lambda)$, gives rise to two *nonreal* zeros of $F_\lambda(z)$ when $\lambda \leq -0.40$. For these values of $\lambda, F_\lambda(z)$ cannot be an element of $\mathcal{L}-\mathcal{P}$, and it follows from (4.11) that -0.40 is a lower bound for the de Bruijn–Newman constant Λ. Actually, based on a somewhat lengthy analysis which utilizes Lemma 9, a better lower bound can be guaranteed. The result, then, of Varga, Norfolk and Ruttan [33] is stated as Theorem 10.

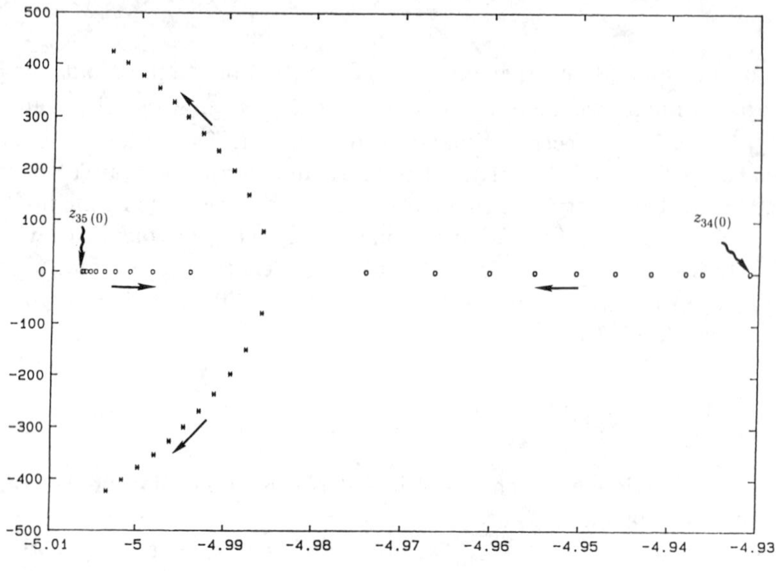

Figure 3.1: $\{z_{34}(-[0.04]j)\}_{j=0}^{20}$ *and* $\{z_{35}(-[0.04]j)\}_{j=0}^{20}$.

THEOREM 10. ([33]). *If* Λ *is the de Bruijn–Newman constant, then*

$$-0.385 < \Lambda. \tag{6.7}$$

The lower bound of Λ in (6.7) is a considerable improvement over the lower bound of (5.5) of Theorem 8, and there is reason to believe that the lower bound of (6.7) *can* be increased further, at the expense of more computation.

What pitfalls in such computations can be foreseen? Note that from (6.5) we have $z_{35}(0) = -50,063.757\cdots$, so that determining $F_\lambda(z)$ accurately from the sum (cf. (4.8))

$$F_\lambda(z) = \sum_{m=0}^{\infty} \frac{\hat{b}_m(\lambda) z^m}{(2m)!},$$

when $z \doteqdot -5 \cdot 10^4$, requires not only adding many terms, but also requires having many accurate moments $\{\hat{b}_m(\lambda)\}_{m=0}^{N}$ available. And this situation only *worsens* as one tracks pairs of successive zeros, say $z_M(0)$ and $z_{M+1}(0)$, when M is large! (For example, when $M = 93$, we have

$$z_{93}(0) = -200,716.54\cdots, \text{ and } z_{94}(0) = -202,469.98\cdots\ .)$$

3.7. Riemann Hypothesis: Necessary and sufficient conditions.

We conclude this chapter with a discussion of some known necessary *and* sufficient conditions for the truth of the Riemann Hypothesis. We begin by stating a known result essentially due to Jensen [16]. (For a complete proof of this, see Csordas and Varga [8].)

THEOREM 11. (Jensen [16]). *Let*

$$f(z) = e^{-\alpha z^2} f_1(z) \qquad (\alpha \geq 0,\ f(z) \not\equiv 0), \tag{7.1}$$

where f_1 *is a real entire function of genus* 0 *or* 1. *Then,* $f \in \mathcal{L}-\mathcal{P}$ *iff*

$$|f'(z)|^2 \geq Re\ \{f(z)\overline{f''(z)}\} \quad (all\ \ z \in \mathbb{C}). \tag{7.2}$$

From (2.6) and the case $\lambda = 0$ of (4.3), we have

$$H_0(x) = \xi(x/2)/8 = \frac{1}{2}\int_{-\infty}^{+\infty} \Phi(t)e^{ixt}dt, \tag{7.3}$$

where $\Phi(t)$ has the properties of (2.5). As we have seen, the Riemann Hypothesis is true iff all the zeros of $H_0(x)$ are real, which is also equivalent to the statement that $H_0 \in \mathcal{L}-\mathcal{P}$. Then, on directly applying the condition of (7.2) to $H_0(z)$ of (7.3), thereby producing double integrals, we have the following known (unpublished) result of Jensen which appeared in Pólya [25].

THEOREM 12. (Pólya [25]). *A necessary and sufficient condition for the truth of the Riemann Hypothesis is that*

$$\Delta(x,y) \geq 0 \ \ for\ all \ \ x,y \in \mathbb{R}, \tag{7.4}$$

where

$$\Delta(x,y) := \int_{-\infty}^{\infty}\int_{-\infty}^{\infty} \Phi(t)\Phi(s)e^{i(t+s)x}e^{(t-s)y}(t-s)^2 dtds. \tag{7.5}$$

From symmetry considerations (cf. (2.5*iii*)), it is sufficient to evaluate $\Delta(x,y)$ of (7.5) only for $x,y \geq 0$, but even with this reduction, checking the condition of (7.4) that $\Delta(x,y) \geq 0$, by means of numerical integration, is at best formidable. However, we next state a recent result of Csordas and Varga [8], using extensions of the Hermite–Biehler Theorem, which further reduces the domain where the positivity $\Delta(x,y)$ needs to be established.

THEOREM 13. ([8]). *A necessary and sufficient condition for the truth of the Riemann Hypothesis is that*

$$\Delta(x,y) \geq 0 \ \ for \ \ 0 < x < \infty \ \ and \ \ 0 \leq y < 1. \tag{7.6}$$

Other known necessary and sufficient conditions for the truth of the Riemann Hypothesis, involving double integrals, can similarly be found in Pólya [25] and Csordas and Varga [9].

REFERENCES

[1] R.P. Boas, **Entire Functions**, Academic Press, Inc., New York, 1954.

[2] N.G. de Bruijn, *The roots of trigonometric integrals*, Duke J. Math., **17** (1950), pp. 197-226.

[3] J.B. Conrey, *At least two fifths of the zeros of the Riemann zeta function are on the critical line*, Bull. Amer. Math. Soc., **20** (1989), pp. 79-81.

[4] G. Csordas, T.S. Norfolk, and R.S. Varga, *The Riemann Hypothesis and the Turán inequalities*, Trans. Amer. Math. Soc., **296** (1986), pp. 521-524.

[5] G. Csordas, T.S. Norfolk, and R.S. Varga, unpublished manuscript.

[6] ———, *A lower bound for the de Bruijn–Newman constant* Λ, Numer. Math., **52** (1988), pp. 483-497.

[7] G. Csordas and R.S. Varga, *Moment inequalities and the Riemann Hypothesis*, Constr. Approx., **4** (1988), pp. 175-198.

[8] ———, *Fourier transforms and the Hermite–Biehler Theorem*, Proc. Amer. Math. Soc., **107** (1989), pp. 645-652.

[9] ———, *Necessary and sufficient conditions and the Riemann Hypothesis*, Adv. in Appl. Math., to appear.

[10] E. Grosswald, *Generalization of a formula of Hayman, and its applications to the study of Riemann's zeta function*, Illinois J. Math., **10** (1966), pp. 9-23.

[11] ———, *Correction and completion of the paper "Generalization of a formula of Hayman,"* Illinois J. Math., **13** (1969), pp. 276-280.

[12] G.H. Hardy, *Sur les zéros de la fonction* $\zeta(s)$ *de Riemann*, C.R. Acad. Sci. Paris, **158** (1914), pp. 1012-1014.

[13] W.K. Hayman, *A generalization of Stirling's formula*, J. Reine Angew. Math., **196** (1956), pp. 67-95.

[14] P. Henrici, **Applied and Computational Complex Analysis**, Vol. 1, John Wiley & Sons, New York, 1974.

[15] ———, **Applied and Computational Complex Analysis**, Vol. 2, John Wiley & Sons, New York, 1977.

[16] J.L.W.V. Jensen, *Recherches sur la théorie des équations*, Acta Math., **36** (1913), pp. 181-195.

[17] E. Laguerre, **Oeuvres**, Vol. 1, Gauthier-Villars, Paris, 1898.

[18] N. Levinson, *More than one third of the zeros of Riemann's zeta-function are on* $\sigma = 1/2$, Adv. in Math., **13** (1974), pp. 383-436.

[19] J. van de Lune, H.J.J. te Riele, and D.T. Winter, *On the zeros of the Riemann zeta function in the critical strip*, IV, Math. Comp., **46** (1986), pp. 667-681.

[20] Yu.V. Matiyasevich, *Yet another machine experiment in support of Riemann's conjecture* (in Russian), Kiebernetika (1982). English translation in Cybernetics, **18** (1983), pp. 705-707.

[21] C.M. Newman, *Fourier transforms with only real zeros*, Proc. Amer. Math. Soc., **61** (1976), pp. 245-251.

[22] A.M. Odlyzko, *The* 10^{20}*th zero of the Riemann zeta function and its neighbors*, preprint, 1989.

[23] A.M. Ostrowski, **Solution of Equations and Systems of Equations**, Academic Press, Inc., New York, 1960.

[24] G. Pólya, *On the zeros of certain trigonometric integrals*, J. London Math. Soc., **1** (1926), pp. 98-99.

[25] ———, *Über die algebraisch-funktionentheoretischen Untersuchungen von J.L.W.V. Jensen*, Kgl. Danske Vid. Sel. Math.-Fys. Medd., **7** (1927), pp. 3-33.

[26] ———, *Über trigonometrische Integralen mit nur reellen Nullstellen*, J. Reine Angew. Math., **158** (1927), pp. 6-18.

[27] G. Pólya and J. Schur, *Über zwei Arten von Faktorenfolgen in der Theorie der algebraischen Gleichungen*, J. Reine Angew. Math., **144** (1914), pp. 89-113.

[28] H.J.J. te Riele, *Table of the First* 15,000 *Zeros of the Riemann Zeta Function to* 28 *Significant Digits and Related Quantities*, Report Number NW 67/79 of the Mathematisch Centrum, Amsterdam, 1979.

[29] H. te Riele, *A new lower bound for the de Bruijn–Newman constant*, Numer. Math., to appear.

[30] B. Riemann, *Über die Anzahl der Primzahlen unter einer gegebenen Grösse*, Monatsh. der Berliner Akad., (1858/60), pp. 671-680. Also in *Gesammelte Mathematische Werke*, 2nd edition, Teubner, Leipzig, 1982, No. VII, pp. 145-153.

[31] J. Stoer and R. Bulirsch, **Introduction to Numerical Analysis**, Springer-Verlag, Inc., Heidelberg, 1980.

[32] E.C. Titchmarsh, **The Theory of the Riemann Zeta-function**, 2nd edition (revised by D.R. Heath-Brown), Oxford University Press, Oxford, 1986.

[33] R.S. Varga, T.S. Norfolk, and A. Ruttan, *A lower bound for the de Bruijn-Newman constant.* II, in preparation.

CHAPTER 4

Asymptotics for the Zeros of the Partial Sums of exp(z)

4.1. Szegö's Theorem and the curves D_∞ and D_n.

With $s_n(z) := \sum_{j=0}^{n} z^j/j!$ $(n \geq 1)$ denoting the familiar partial sum of the exponential function e^z, we investigate here the location of the zeros of the *normalized* partial sums, $s_n(nz)$, and the rate at which these zeros tend to the Szegö curve D_∞, defined by

$$D_\infty := \{z \in \mathbb{C} : |ze^{1-z}| = 1 \text{ and } |z| \leq 1\}. \tag{1.1}$$

By way of review, the well-known Eneström–Kakeya Theorem (cf. Marden [7, p. 137, Ex. 2]) asserts that, for any polynomial $p_n(z) = \sum_{j=0}^{n} a_j z^j$ with $a_j > 0$ $(0 \leq j \leq n)$, all the zeros of $p_n(z)$ necessarily lie in the closed annulus

$$\min_{0 \leq i < n} \left(\frac{a_i}{a_{i+1}}\right) \leq |z| \leq \max_{0 \leq i < n} \left(\frac{a_i}{a_{i+1}}\right).$$

Applying the final inequality above to the partial sum $s_n(z)$ of e^z immediately shows that $s_n(z)$ has all its zeros in $|z| \leq n$, for every $n \geq 1$. A sharpened form of the Eneström–Kakeya Theorem (cf. Anderson , Saff , and Varga [1, Cor. 2]) actually shows that all zeros of $s_n(z)$ satisfy $|z| < n$ for any $n > 1$. Thus, if $\{z_{k,n}\}_{k=1}^{n}$ is the set of the zeros of the *normalized* partial sum $s_n(nz)$, then these zeros lie in the closed unit disk $\Delta := \{z \in \mathbb{C} : |z| \leq 1\}$ for every $n \geq 1$, and they lie in the *interior* of Δ for every $n > 1$. (This can be seen quite clearly in Figure 4.1.) Consequently, the infinite set of all zeros of all normalized partial sums $\{s_n(nz)\}_{n=1}^{\infty}$ must have at *least* one accumulation point in the compact set Δ.

In his remarkable paper of 1924, Szegö [12] established the following theorem.

THEOREM 1. (Szegö [12]). *Each accumulation point z in Δ of the zeros of the normalized partial sums $\{s_n(nz)\}_{n=1}^{\infty}$ of e^z must lie on the curve D_∞ of* (1.1). *Conversely, each point of D_∞ is an accumulation point of the zeros of the normalized partial sums $\{s_n(nz)\}_{n=1}^{\infty}$ of e^z.*

Subsequently, it was shown by Buckholtz [2] that the zeros of $s_n(nz)$ lie *outside* the curve D_∞, for every $n \geq 1$. To indicate these results, we have graphed in Figure 4.2 the 16 zeros of $s_{16}(16\,z)$ (these zeros being represented by $\times$'s), along with the Szegö curve D_∞ (cf. (1.1)) and $\partial\Delta$, the boundary of Δ. The same is done in Figure 4.3 with the 27 zeros of $s_{27}(27\,z)$.

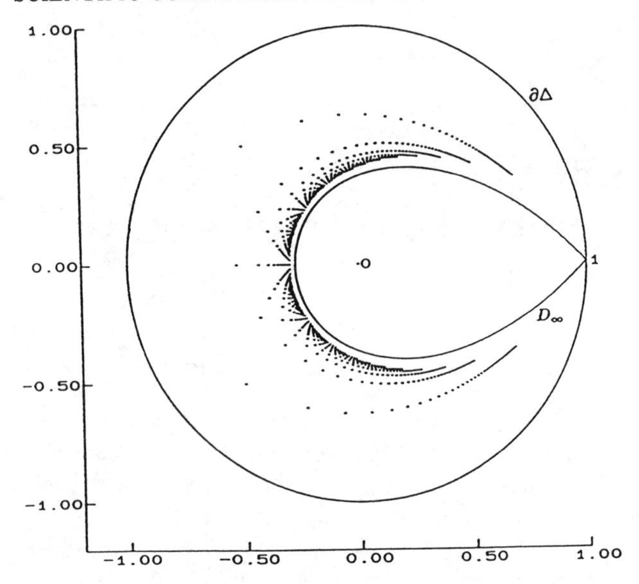

Figure 4.1: *Zeros of* $\{s_n(nz)\}_{n=1}^{40}$.

Figures 4.1, 4.2, and 4.3 indicate that the zeros of $s_n(nz)$ converge in a seemingly "regular" way to the curve D_∞, and these figures also indicate that this convergence seems *slowest* in a neighborhood of the point $z = 1$ of D_∞. Because we are interested in measuring the *slowest* rate of convergence of these zeros to the curve D_∞, let dist $[z; B] := \inf\{|z - w| : w \in B\}$ denote the usual (metric) distance between a point z and a set B in the complex plane, and, for a finite set $\{t_j\}_{j=1}^m$, we define

$$\text{(1.2)} \qquad \text{dist } \left[\{t_j\}_{j=1}^m; B\right] := \max_{1 \leq j \leq m} (\text{dist } [t_j; B]) .$$

Buckholtz [2] established the result that the zeros $\{z_{k,n}\}_{k=1}^n$ of $s_n(nz)$ all lie within a distance of $\frac{2e}{\sqrt{n}}$ from D_∞, i.e., in the notation of (1.2),

$$\text{(1.3)} \qquad \text{dist } \left[\{z_{k,n}\}_{k=1}^n ; D_\infty\right] \leq \frac{2e}{\sqrt{n}} \qquad (n = 1, 2, \cdots).$$

This implies of course that

$$\text{(1.3}'\text{)} \qquad \overline{\lim_{n \to \infty}} \left\{\sqrt{n} \cdot \text{dist } [\{z_{k,n}\}_{k=1}^n; D_\infty]\right\} \leq 2e = 5.43656\cdots .$$

What has not been considered in the literature is whether the upper bound of (1.3) is *best possible* as a function of n, as $n \to \infty$. Based on the results of Newman and Rivlin [8] and Saff and Varga [10], it can be easily deduced that the upper bound of (1.3), as a function of n, is indeed *best possible*. More precisely, from Carpenter, Varga, and Waldvogel [4], we have the following result (whose proof is given in §4.2).

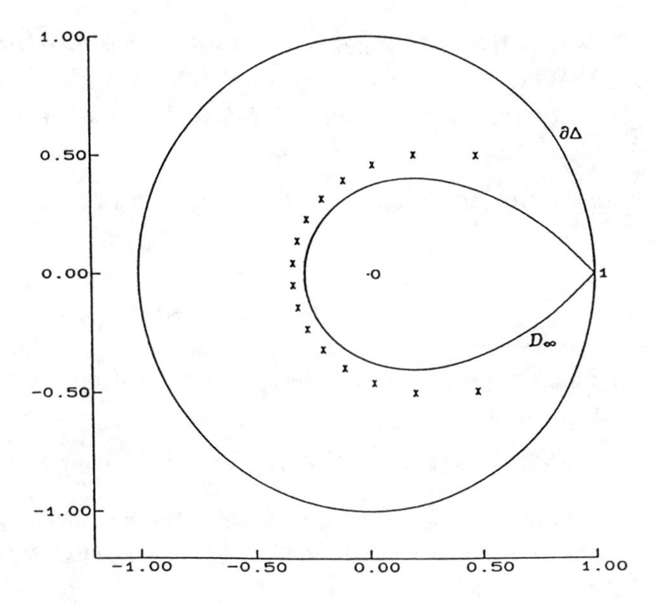

Figure 4.2: *Zeros of* $s_{16}(16z)$.

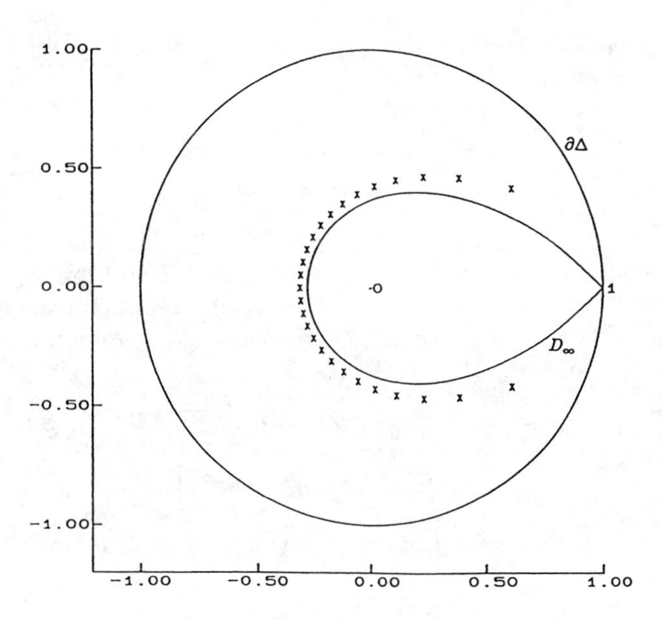

Figure 4.3: *Zeros of* $s_{27}(27z)$.

PROPOSITION 2. ([4]). *If* $\{z_{k,n}\}_{k=1}^{n}$ *denotes the zeros of* $s_n(nz)$ *and if* t_1 *denotes* (cf. (2.2)) *the zero of* $erfc(w) := \frac{2}{\sqrt{\pi}} \int_w^\infty e^{-t^2}\, dt$ *with* $Im\ t_1 > 0$ *which is closest to the origin, then the result of* (1.3) *is best possible as a function of* n, *in that*

$$\varliminf_{n\to\infty} \{\sqrt{n} \cdot \text{dist } [\{z_{k,n}\}_{k=1}^{n}; D_\infty]\} \geq (Im\ t_1 + \ Re\ t_1) = 0.63665\cdots . \tag{1.4}$$

On examining Figures 4.2 and 4.3, we note that there is apparently *faster convergence* (to the curve D_∞) of those zeros of $\{z_{k,n}\}_{k=1}^{n}$ which stay uniformly away from the point $z = 1$. In fact, if we use the open disk C_δ about the point $z = 1$, i.e.,

$$C_\delta := \{z \in \mathbb{C} : |z - 1| < \delta\} \qquad (0 < \delta \leq 1), \tag{1.5}$$

to *exclude* points of $\{z_{k,n}\}_{k=1}^{n}$ near $z = 1$, this observed faster convergence can be *quantified.* More precisely, we shall prove in §4.2 the new result of the following theorem.

THEOREM 3. ([4]). *If* $\{z_{k,n}\}_{k=1}^{n}$ *denotes the zeros of the normalized partial sums* $s_n(nz)$ *and if* δ *is any fixed number with* $0 < \delta \leq 1$, *then*

$$\text{dist } [\{z_{k,n}\}_{k=1}^{n} \backslash C_\delta; D_\infty] = O\left(\frac{\log\, n}{n}\right) \qquad (n \to \infty). \tag{1.6}$$

Moreover, the bound $O((\log\, n)/n)$ *of* (1.6) *is best possible as a function of* n, *in the sense that*

$$\varliminf_{n\to\infty} \left\{\frac{n}{\log n} \cdot \text{dist } \left[\{z_{k,n}\}_{k=1}^{n} \backslash C_\delta; D_\infty\right]\right\} \geq 0.10900\cdots > 0, \tag{1.6$'$}$$

for any $0 < \delta \leq 1$.

The seemingly "regular" way the zeros $\{z_{k,n}\}_{k=1}^{n}$ of $s_n(nz)$ are distributed in Figures 4.2 and 4.3, suggests that there may be a smooth arc D_n (dependent on n) in the unit disk Δ which provides a much closer approximation to the zeros $\{z_{k,n}\}_{k=1}^{n}$ of $s_n(nz)$, than does the curve D_∞. As already suggested from the work of Szegö [12], we set

$$\begin{aligned} D_n := \ & \left\{z \in \mathbb{C} : |ze^{1-z}|^n = \tau_n \sqrt{2\pi n} \left|\tfrac{1-z}{z}\right|, |z| \leq 1, \right. \\ & \left. \text{and } |\arg\, z| \geq \cos^{-1}\left(\tfrac{n-2}{n}\right)\right\}, \end{aligned} \tag{1.7}$$

for each $n \geq 1$, where from Stirling's asymptotic series formula (cf. Henrici [6, p. 377]),

$$\begin{aligned} \tau_n := \frac{n!}{n^n e^{-n}\sqrt{2\pi n}} \approx \ & 1 + \frac{1}{12n} + \frac{1}{288n^2} \\ & - \frac{139}{51\,840n^3} + \cdots \quad (n \to \infty). \end{aligned} \tag{1.8}$$

We mention that $\log \tau_n$ can be expressed (cf. Henrici [6, p. 377]) in terms of the Binet function $J(z)$, which has the following asymptotic series representation (cf. [6, p. 359]):

$$\log \tau_n = J(n) \approx \frac{n^{-1}}{12} - \frac{n^{-3}}{360} + \frac{n^{-5}}{1260} - \frac{n^{-7}}{1680} + \frac{n^{-9}}{1188} - \cdots \qquad (n \to \infty).$$

Further, the arc D_n of (1.7) can be shown (cf. [4, Prop. 3]) to be a well-defined curve which lies interior to the closed unit disk Δ for each positive integer n. It is also interesting to point out that (cf. [4, Prop. 3]) the arc D_n is, for each n, *star-shaped* with respect to the origin, i.e., for any fixed real number θ with $|\theta| \geq \cos^{-1}\left(\frac{n-2}{n}\right)$, the ray $\{z = re^{i\theta} : r \geq 0\}$ intersects D_n in a unique point. We further mention that the restriction that $|\arg z| \geq \cos^{-1}\left(\frac{n-2}{n}\right)$, in the definition of D_n in (1.7), comes from the fact that $s_n(nz)$ has all its zeros in the sector $|\arg z| \geq \cos^{-1}\left(\frac{n-2}{n}\right)$ (cf. Saff and Varga [9]).

With the arc D_n of (1.7), we present the following new result, whose proof appears in §4.3.

THEOREM 4. ([4]). *If* $\{z_{k,n}\}_{k=1}^{n}$ *denotes the zeros of the normalized partial sums* $s_n(nz)$ *and if* δ *is any fixed number with* $0 < \delta \leq 1$, *then*

$$\text{(1.9)} \qquad \operatorname{dist}\,[\{z_{k,n}\}_{k=1}^{n} \backslash C_\delta; D_n] = O\left(\frac{1}{n^2}\right) \qquad (n \to \infty).$$

Moreover, the bound $O\left(\frac{1}{n^2}\right)$ *of* (1.9) *is best possible as a function of* n, *in the sense that*

$$\text{(1.9}'\text{)} \qquad \varliminf_{n\to\infty} \left\{n^2 \cdot \operatorname{dist}\,\left[\{z_{k,n}\}_{k=1}^{n} \backslash C_\delta; D_n\right]\right\} \geq 0.13326\cdots > 0,$$

for any $0 < \delta \leq 1$.

To illustrate the result of Theorem 4, we have graphed the 16 zeros of $s_{16}(16z)$, along with the curve D_{16}, in Figure 4.4. The same is done in Figure 4.5 for the 27 zeros of $s_{27}(27z)$ and the curve D_{27}. Up to plotting accuracy, it appears that the zeros of $s_{16}(16z)$ and $s_{27}(27z)$ lie, respectively, *on* the curves D_{16} and D_{27}!

4.2. Proofs of Proposition 2 and Theorem 3.

We begin with the following proof.

Proof of Proposition 2. As shown in Newman and Rivlin [8],

$$\text{(2.1)} \qquad \left\{\frac{s_n\left(n + \sqrt{2n}\,w\right)}{e^{n+\sqrt{2n}\,w}}\right\}_{n=1}^{\infty} \text{ converges uniformly to } \frac{1}{\sqrt{\pi}}\int_w^\infty e^{-t^2}dt =: \frac{1}{2}\operatorname{erfc}(w),$$

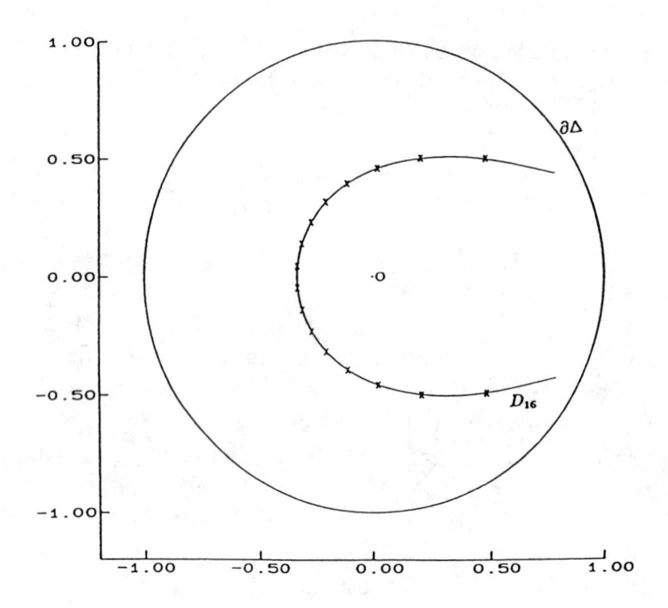

Figure 4.4: D_{16} *and the zeros of* $s_{16}(16z)$.

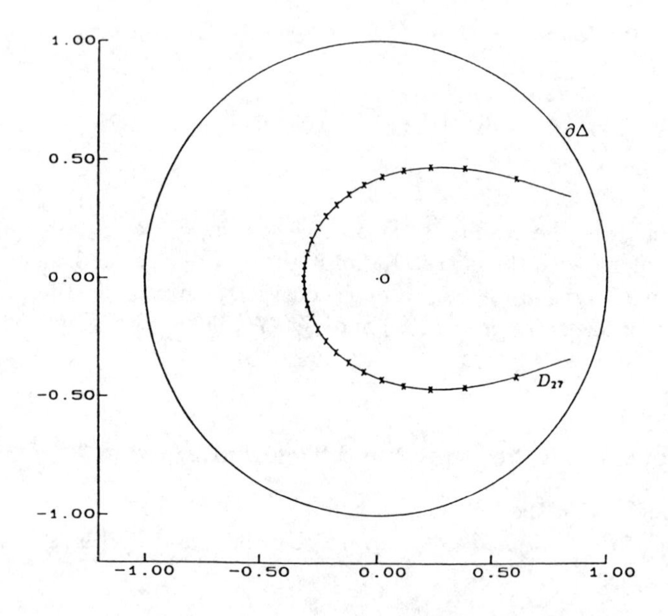

Figure 4.5: D_{27} *and the zeros of* $s_{27}(27z)$.

as $n \to \infty$, on any compact set in the upper-half complex plane (i.e., $Im\ w \geq 0$). If t_1 is the zero of erfc (w) in the upper-half plane which is closest to the origin, then it is known numerically (cf. Fettis, Caslin, and Cramer [5]) that

$$t_1 = -1.35481 \cdots + i1.99146 \cdots . \tag{2.2}$$

Now, as the only zeros of $s_n(n + \sqrt{2n}\, w)e^{-n-\sqrt{2n}\, w}$ are values of w for which $s_n(n+\sqrt{2n}\, w)$ vanishes, then the uniform convergence in (2.1) implies, with Hurwitz's Theorem (cf. Titchmarsh [13, p. 119]), that $s_n(n + \sqrt{2n}\, w)$ has a zero in any small closed disk with center t_1, for all n sufficiently large. Consequently, as shown in Saff and Varga [10], $s_n(n + \sqrt{2n}\, w)$ has a zero of the form

$$n + \sqrt{2n}\,(t_1 + o(1)) = n\left\{1 + \sqrt{\frac{2}{n}}\,(t_1 + o(1))\right\} \qquad (n \to \infty),$$

or equivalently, $s_n(nz)$ has a zero $z_{1,n}$ of the form

$$z_{1,n} := 1 + \sqrt{\frac{2}{n}}\,(t_1 + o(1)) \qquad (n \to \infty). \tag{2.3}$$

Next, if z lies on D_∞ with $Re\ z = 1 - \delta$ where $\delta > 0$ is small, it directly follows from the definition of D_∞ in (1.1) that $(1-\delta)^2 + (Im\ z)^2 = e^{-2\delta}$, so that

$$|Im\ z| = \delta\left\{1 - \frac{2}{3}\,\delta + O(\delta^2)\right\} \qquad (\delta \to 0). \tag{2.4}$$

Note that (2.4) establishes that the curve D_∞, in the upper-half plane, makes an angle of $\pi/4$ with the real axis as it passes through $z = 1$. (This can also be seen from Figures 4.2 and 4.3.) From (2.3) and (2.4), a calculation shows (cf. Carpenter [3, p. 137]) that, for n large, the distance of $z_{1,n}$ of (2.3) to the curve D_∞ satisfies

$$\text{dist}\ [z_{1,n}; D_\infty] = \frac{1}{\sqrt{n}}\{Im\ t_1 + \ Re\ t_1 + o(1)\} \qquad (n \to \infty),$$

whence

$$\lim_{n\to\infty}\{\sqrt{n}\ \text{dist}\ [z_{1,n}; D_\infty]\} = Im\ t_1 + \ Re\ t_1 = 0.63665\cdots, \tag{2.5}$$

the last result utilizing the numerical estimates of (2.2). But from (1.2), as dist $[z_{1,n}; D_\infty] \leq$ dist $\left[\{z_{k,n}\}_{k=1}^{n}; D_\infty\right]$, we have the desired result of (1.4). $\square$

For the proof of Theorem 3 of §4.1, we need the following construction. From the definition $s_n(z) := \sum_{j=0}^{n} z^j/j!$, differentiation establishes that

$$e^{-z}s_n(z) = 1 - \frac{1}{n!}\int_0^z \zeta^n e^{-\zeta}\, d\zeta. \tag{2.6}$$

Replacing ζ by $n\zeta$ and z by nz in the above expression results in

$$e^{-nz}s_n(nz) = 1 - \frac{n^{n+1}}{n!}\int_0^z \zeta^n e^{-n\zeta}\,d\zeta. \tag{2.7}$$

Using the definition of τ_n of (1.8) in (2.7) then gives

$$e^{-nz}s_n(nz) = 1 - \frac{\sqrt{n}}{\tau_n\sqrt{2\pi}}\int_0^z \left(\zeta e^{1-\zeta}\right)^n d\zeta. \tag{2.8}$$

Next, from Szegö [12], it is known that $w = \zeta e^{1-\zeta}$ is *univalent* in $|\zeta| < 1$. (For a proof of this, see the special case $\sigma = 0$ in Saff and Varga [11, Lem. 4.1].) Since we are ultimately interested only in the zeros of $s_n(nz)$ (which, from §4.1, must lie in $|z| < 1$ for all $n > 1$), we make the change of variables $w = \zeta e^{1-\zeta}$ in (2.8), which gives

$$e^{-nz}s_n(nz) = 1 - \frac{\sqrt{n}}{\tau_n\sqrt{2\pi}}\int_0^{ze^{1-z}} w^{n-1}\left(\frac{\zeta(w)}{1-\zeta(w)}\right)dw. \tag{2.9}$$

The form of the above integral brings us to the following result, which is again motivated by the original work of Szegö [12]. Consider the integral

$$\int_0^A w^{n-1}G(w)dw, \tag{2.10}$$

where the path of integration in (2.10) is taken to be the complex line segment joining 0 and A. Assuming that $G(w)$ is analytic in an open region containing this line segment $[0, A]$, then expanding $G(w)$ in a Taylor's series about the point $w = A$ gives

$$\int_0^A w^{n-1}G(w)dw = \sum_{m=0}^{\infty}\frac{G^{(m)}(A)}{m!}\int_0^A w^{n-1}(w-A)^m dw. \tag{2.11}$$

Since the integral associated with the mth term of the above sum is (after a change of variables) just the beta integral (cf. [6, p. 55]), this term then satisfies

$$\frac{G^{(m)}(A)}{m!}\int_0^A w^{n-1}(w-A)^m dw = \frac{(-1)^m A^{m+n}G^{(m)}(A)}{\prod_{j=0}^{m}(n+j)},$$

for each $m \geq 0$ and $n \geq 1$. Thus, the integral of (2.10) can be expressed as

$$\int_0^A w^{n-1}G(w)dw = A^n\sum_{m=0}^{\infty}\frac{(-1)^m A^m G^{(m)}(A)}{\prod_{j=0}^{m}(n+j)}. \tag{2.12}$$

We now connect the two integrals of (2.9) and (2.10) by setting $A := ze^{1-z}$, $F(\zeta) := \zeta/(1-\zeta)$, and $G(w) := F(\zeta(w))$, where $w := \zeta e^{1-\zeta}$. If z is any *interior point* of the closed unit disk Δ, then $G(w)$, so defined, is indeed analytic in an open region containing the line segment $[0, ze^{1-z}]$, and the representation of (2.12) is valid. A short calculation of the explicit values of $G^{(m)}(ze^{1-z})$, for $0 \leq m \leq 4$, allows us in this case to give the first few terms of (2.12):

$$\int_0^{ze^{1-z}} w^{n-1}\left(\frac{\zeta(w)}{1-\zeta(w)}\right) dw = \frac{z(ze^{1-z})^n}{n(1-z)} \cdot$$

$$\left\{1 - \frac{1}{(n+1)(1-z)^2} + \frac{z(4-z)}{(n+1)(n+2)(1-z)^4} - \frac{z^2(27-14z+2z^2)}{(n+1)(n+2)(n+3)(1-z)^6} + \frac{z^3(256-203z+58z^2-6z^3)}{(\prod_{j=1}^{4}(n+j))(1-z)^8} - \cdots\right\} \quad (n = 1, 2, \cdots). \tag{2.13}$$

Next, estimating the Cauchy remainder for the sections of the Taylor series in (2.11) shows, with (2.12), that, for each nonnegative integer N,

$$\int_0^{ze^{1-z}} w^{n-1}\left(\frac{\zeta(w)}{1-\zeta(w)}\right) dw = \left(ze^{1-z}\right)^n \sum_{m=0}^{N} \frac{(-1)^m (ze^{1-z})^m G^{(m)}(ze^{1-z})}{\prod_{j=0}^{m}(n+j)} + O\left(\frac{1}{n^{N+2}}\right) \quad (n \to \infty), \tag{2.14}$$

uniformly on any compact subset of $\Delta\backslash\{1\}$. (The motivation for this result of course comes from Szegö [12], who derived (2.14) for the case $N = 0$.)

As a consequence of the case $N = 0$ in (2.14) and (2.13), we have

$$\int_0^{ze^{1-z}} w^{n-1}\left(\frac{\zeta(w)}{1-\zeta(w)}\right) dw = \frac{z(ze^{1-z})^n}{n(1-z)}\left\{1 + O\left(\frac{1}{n}\right)\right\} \quad (n \to \infty), \tag{2.15}$$

uniformly on any compact subset of $\Delta\backslash\{1\}$. Thus, if z is a zero of the normalized partial sum $s_n(nz)$ of e^z, then from (2.9) and (2.15) we have

$$1 - \frac{\sqrt{n}}{\tau_n\sqrt{2\pi}} \cdot \frac{z(ze^{1-z})^n}{n(1-z)}\left\{1 + O\left(\frac{1}{n}\right)\right\} = 0,$$

or equivalently,

$$(ze^{1-z})^n = \tau_n\sqrt{2\pi n}\left(\frac{1-z}{z}\right)\left\{1 + O\left(\frac{1}{n}\right)\right\} (n \to \infty), \tag{2.16}$$

uniformly on any compact subset of $\Delta\backslash\{1\}$.

Now, we arrive at the following proof.

Proof of Theorem 3. From Szegö [12], it is known that $w = ze^{1-z}$ maps the interior of the Szegö curve, D_∞, conformally onto the interior of $|w| < 1$, and it also maps, in a 1-1 fashion, the points of D_∞ onto $|w| = 1$, such that the argument of $w = ze^{1-z}$, as z traverses D_∞ in the positive sense starting at $z = 1$, increases monotonically from 0 to 2π. Szegö [12] also showed that the zeros of $s_n(nz)$ are asymptotically *uniformly distributed* in angle under the mapping $w = ze^{1-z}$ (as $n \to \infty$). More precisely, let ϕ_1 and ϕ_2 be any real numbers with $0 < \phi_1 < \phi_2 < 2\pi$, and let z_1 and z_2 be, respectively, the inverse images of $w_1 = e^{i\phi_1}$ and $w_2 = e^{i\phi_2}$ under the mapping $w = ze^{1-z}$, so that z_1 and z_2 are points of D_∞, with $0 < \arg z_1 < \arg z_2 < 2\pi$. Let S be the sector in the z-plane, defined by

$$S = \{z \in \mathbb{C} : \arg z_1 \leq \arg z \leq \arg z_2\}.$$

Then, if σ_n is the number of zeros of $s_n(nz)$ in S, Szegö [12] showed that

$$\lim_{n\to\infty} \frac{\sigma_n}{n} = \frac{\phi_2 - \phi_1}{2\pi}.$$

This result implies that, for n large, the zeros of $s_n(nz)$ are roughly *uniformly distributed* in angle in the w-plane, under the mapping $w = ze^{1-z}$.

This can be used as follows. If we take the n uniformly distributed points $\{\exp[i(2k-1)\pi/n]\}_{k=1}^n$ on $|w| = 1$, let $\{\tilde{z}_{k,n}\}_{k=1}^n$ be the unique inverse images of these points in the z-plane under the mapping $w = ze^{1-z}$, i.e.,

$$\tilde{z}_{k,n}e^{1-\tilde{z}_{k,n}} = \exp[i(2k-1)\pi/n] \quad (k = 1, 2, \cdots, n). \tag{2.17}$$

By definition, the points $\{\tilde{z}_{k,n}\}_{k=1}^n$ lie on D_∞, and we have graphed in Figure 4.6 the points $\{\tilde{z}_{k,16}\}_{k=1}^{16}$ as $*$'s on D_∞, along with the zeros $\{z_{k,16}\}_{k=1}^{16}$ of $s_{16}(16z)$. The same is done in Figure 4.7 for $\{\tilde{z}_{k,27}\}_{k=1}^{27}$ and the zeros $\{z_{k,27}\}_{k=1}^{27}$ of $s_{27}(27z)$.

From (2.17), we have

$$\left(\tilde{z}_{k,n}e^{1-\tilde{z}_{k,n}}\right)^n = -1 \qquad (k = 1, 2, \cdots, n). \tag{2.18}$$

Regarding $\tilde{z}_{k,n}$ as an approximation of $z_{k,n}$, write $z_{k,n}$, a zero of $s_n(nz)$, as $z_{k,n} = \tilde{z}_{k,n} + \delta_{k,n}$, and insert this in (2.16). On using (2.18), a straightforward calculation shows, on taking logarithms and dividing by n, that

$$\begin{aligned} &-\left(1 - \tfrac{1}{\tilde{z}_{k,n}}\right)\delta_{k,n} + O(\delta_{k,n}^2) \\ &= \tfrac{1}{n}\log\left\{\tau_n\sqrt{2\pi\, n}\left(1 - \tfrac{1}{\tilde{z}_{k,n}}\right)\right\} + O\left(\tfrac{\delta_{k,n}}{n}\right) + O\left(\tfrac{1}{n^2}\right), \end{aligned} \tag{2.19}$$

provided that we consider only those zeros $z_{k,n}$ which lie outside of C_δ (where δ is a fixed number with $0 < \delta \leq 1$ and where C_δ is defined in (1.5)). (For

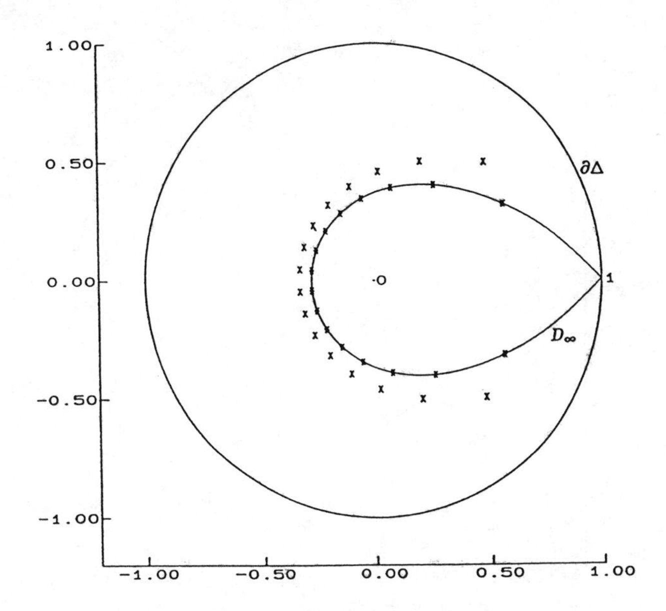

Figure 4.6: $\{z_{k,16}\}_{k=1}^{16}$ *and* $\{\tilde{z}_{k,n}\}_{k=1}^{n}$.

the logarithm in (2.19), we choose its single-valued extension on $\mathbb{C}\backslash[0,+\infty)$ for which $\log(-1) = i\pi$.)

For the set $\{z_{k,n}\}_{k=1}^{n}\backslash C_\delta$, there evidently exists a positive constant c_1, dependent only on δ, such that

$$0 < c_1 \leq \left|1 - \frac{1}{\tilde{z}_{k,n}}\right|,$$

for all points $\tilde{z}_{k,n}$ associated with points of $\{z_{k,n}\}_{k=1}^{n}\backslash C_\delta$. Thus, we deduce from (2.19) that

$$\text{(2.20)} \qquad \delta_{k,n} = O\left(\frac{\log \tau_n \sqrt{2\pi}\, n}{n}\right) = O\left(\frac{\log n}{n}\right) \qquad (n \to \infty),$$

for all points of $\{z_{k,n}\}_{k=1}^{n}\backslash C_\delta$. But because $\tilde{z}_{k,n}$ is not necessarily the closest point of D_∞ to $z_{k,n}$, it follows that dist $[z_{k,n}; D_\infty] \leq |\delta_{k,n}|$ for all points of $\{z_{k,n}\}_{k=1}^{n}\backslash C_\delta$, which from (2.20) gives the desired result of (1.6) of Theorem 3.

To show that the result of (1.6) of Theorem 3 is *best possible* as a function of n, first let $n = 2m+1$ be any odd positive integer, and let $-\mu$ denote the *negative* real point of the Szegö curve D_∞, i.e., μ is the unique positive root of $\mu e^{1+\mu} = 1$ and numerically, $\mu = 0.27846\cdots$. From (2.18), we see that

$$\text{(2.21)} \qquad \tilde{z}_{m+1,2m+1} = -\mu = -\ 0.27846\cdots \qquad (m = 0, 1, \cdots),$$

and it similarly follows that $z_{m+1,2m+1}$ is the unique real (negative) zero of the odd polynomial $s_{2m+1}((2m+1)z)$. Using (2.21), it follows from (2.19)

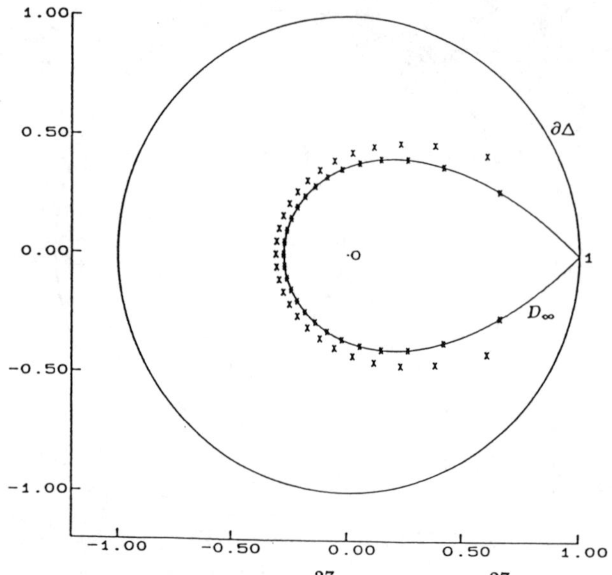

Figure 4.7: $\{z_{k,27}\}_{k=1}^{27}$ *and* $\{\tilde{z}_{k,27}\}_{k=1}^{27}$.

that

$$\lim_{m\to\infty}\left\{\left(\frac{2m+1}{\log(2m+1)}\right)\cdot\delta_{m+1,2m+1}\right\} = -\frac{1}{2(1+\frac{1}{\mu})} \tag{2.22}$$

$$= -\ 0.10890\cdots.$$

Since $z_{m+1,2m+1}$ lies *outside* the curve D_∞, note that $\delta_{m+1,2m+1} < 0$, that $-\delta_{m+1,2m+1} = \text{dist}\ [z_{m+1,2m+1}; D_\infty]$, and that $z_{m+1,2m+1}$ lies outside C_δ for *any* δ with $0 < \delta \le 1$. Thus, (2.22) becomes

$$\lim_{m\to\infty}\left\{\left(\frac{2m+1}{\log(2m+1)}\right)\cdot\ \text{dist}\ [z_{m+1,2m+1}; D_\infty]\right\} \tag{2.23}$$

$$= \frac{1}{2(1+\frac{1}{\mu})} = 0.10890\cdots.$$

But since $\text{dist}\ [z_{m+1,2m+1}; D_\infty] \le \text{dist}\ \left[\{z_{k,2m+1}\}_{k=1}^{2m+1}\backslash C_\delta; D_\infty\right]$, then from (2.23),

$$\varliminf_{m\to\infty}\left\{\left(\frac{2m+1}{\log(2m+1)}\right)\cdot\text{dist}\ \left[\{z_{k,2m+1}\}_{k=1}^{2m+1}\backslash C_\delta; D_\infty\right]\right\} \tag{2.24}$$

$$\ge \frac{1}{2(1+\frac{1}{\mu})} = 0.10890\cdots > 0.$$

For the case when $n = 2m$ is an even positive integer, it can be similarly shown that the analogue of (2.23) is

$$\text{(2.25)} \quad \lim_{m\to\infty}\left\{\frac{2m}{\log(2m)}\cdot \text{ dist } [z_{m,2m}; D_\infty]\right\} = \frac{1}{2(1+\frac{1}{\mu})} = 0.10890\cdots,$$

so that (cf. (2.24))

$$\text{(2.26)} \quad \underline{\lim}_{m\to\infty}\left\{\frac{2m}{\log(2m)}\cdot \text{ dist } \left[\{z_{k,2m}\}_{k=1}^{2m}\backslash C_\delta; D_\infty\right]\right\} \geq \frac{1}{2(1+\frac{1}{\mu})} > 0.$$

Combining (2.24) and (2.25) gives

$$\text{(2.27)} \quad \left\{\frac{n}{\log n}\cdot \text{ dist } [\{z_{k,n}\}_{k=1}^{n}\backslash C_\delta; D_\infty]\right\} \geq 0.10890\cdots > 0,$$

which is the desired result of (1.6′) of Theorem 3. □

4.3. Proof of Theorem 4.

In (2.16), we have a relationship which holds, uniformly in n as $n \to \infty$, for any zero z of $s_n(nz)$ which lies in a compact subset of $\Delta\backslash\{1\}$. On taking moduli and nth roots in (2.16), it is eminently clear *why*, as $n \to \infty$, the Szegö curve, D_∞ of (1.1), emerges as the *only* possible place where the zeros of $\{s_n(nz)\}_{n=1}^{\infty}$ can asymptotically accumulate. As we know from Proposition 2, the maximum distance of the zeros of $s_n(nz)$ to the curve D_∞ is $O(1/\sqrt{n})$ as $n \to \infty$, and this maximum distance can be improved in Theorem 3 to $O((\log n)/n)$ on compact subsets of $\Delta\backslash\{1\}$.

But, it is natural to ask if there is a way of defining an arc, say, D_n, now depending on n, for which the zeros of $s_n(nz)$ lie *substantially* closer to D_n than to the curve D_∞. Of course, any smooth curve through the zeros $\{z_{k,n}\}_{k=1}^{n}$ of $s_n(nz)$ would affirmatively answer this question. It turns out that it is possible to *define* such an arc D_n, without explicit knowledge of the zeros $\{z_{k,n}\}_{k=1}^{n}$ of $s_n(nz)$. In fact, one obtains the definition of the arc D_n in (1.7) by dropping the term $O\left(\frac{1}{n}\right)$ in (2.16) and taking moduli throughout!

Now, we arrive at the following proof.

Proof of Theorem 4. For any fixed δ with $0 < \delta \leq 1$, we consider the set $\{z_{k,n}\}_{k=1}^{n}\backslash C_\delta$, where C_δ is defined in (1.5) and where $\{z_{k,n}\}_{k=1}^{n}$ is again the set of zeros of $s_n(nz)$. Then, for any zeros $z_{k,n}$ in $\{z_{k,n}\}_{k=1}^{n}\backslash C_\delta$, (2.16) is valid, i.e.,

$$\text{(3.1)} \quad (ze^{1-z})^n = \tau_n\sqrt{2\pi n}\left(\frac{1-z}{z}\right)\left\{1+O\left(\frac{1}{n}\right)\right\} \quad (n\to\infty),$$

where the constant implicit in $O(\frac{1}{n})$ depends only on δ. On the other hand, from our remarks concerning the arc D_n, we have, for any θ with $\theta_n \leq \theta \leq 2\pi - \theta_n$, where $\theta_n := \cos^{-1}((n-2)/n)$ for each $n \geq 1$, that there is a unique

$r_n(\theta)$ in $(0,1)$ such that $z = r_n(\theta)e^{i\theta}$ lies on the curve D_n. This implies from (1.7) that there is a real number $\Psi(n,\theta)$ such that

$$\frac{z(ze^{1-z})^n}{\tau_n\sqrt{2\pi n}(1-z)} = e^{i\Psi(n,\theta)}, \tag{3.2}$$

where

$$\Psi(n,\theta) := n[\theta - r_n(\theta)\sin\theta] + \theta + \tan^{-1}\left[\frac{r_n(\theta)\sin\theta}{1 - r_n(\theta)\cos\theta}\right], \tag{3.3}$$

for all $\theta_n \le \theta \le 2\pi - \theta_n$. It turns out that $\Psi(n,\theta)$ is a strictly increasing function of θ on $[\theta_n, 2\pi - \theta_n]$, and that there are exactly n distinct values of θ in $(\theta_n, 2\pi - \theta_n)$ for which $\Psi(n,\theta) \equiv 0 (\mathrm{mod}\ 2\pi)$. If we denote these n particular points on D_n by $\{\hat{z}_{k,n}\}_{k=1}^n$, it follows from (3.2) that

$$\left(\hat{z}_{k,n}e^{1-\hat{z}_{k,n}}\right)^n = \tau_n\sqrt{2\pi\, n}\left(\frac{1-\hat{z}_{k,n}}{\hat{z}_{k,n}}\right) \qquad (k = 1, 2, \cdots, n). \tag{3.4}$$

Recalling that $\{z_{k,n}\}_{k=1}^n$ is the set of zeros (with increasing arguments) of $s_n(nz)$, express $z_{k,n}$ as $z_{k,n} = \hat{z}_{k,n} + \delta_{k,n}$. Thus, for any zero $z_{k,n}$ in $\{z_{k,n}\}_{k=1}^n \backslash C_\delta$, we have from (3.1) that

$$\left(z_{k,n}e^{1-z_{k,n}}\right)^n = \tau_n\sqrt{2\pi\, n}\left(\frac{1-z_{k,n}}{z_{k,n}}\right)\left\{1 + O\left(\frac{1}{n}\right)\right\}. \tag{3.5}$$

Replacing $z_{k,n}$ by $\hat{z}_{k,n} + \delta_{k,n}$ in (3.5) and using (3.4), this reduces, when logarithms are taken and when a division by n is made, to

$$\log\left(1 + \frac{\delta_{k,n}}{\hat{z}_{k,n}}\right) - \delta_{k,n} = \frac{1}{n}\log\left(1 - \frac{\delta_{k,n}}{1-\hat{z}_{k,n}}\right) - \frac{1}{n}\log\left(1 + \frac{\delta_{k,n}}{\hat{z}_{k,n}}\right) + O\left(\frac{1}{n^2}\right), \tag{3.6}$$

as $n \to \infty$. On expanding these various terms (with the assumption that $\delta_{k,n}$ is sufficiently small), one easily determines that

$$\delta_{k,n} = O\left(\frac{1}{n^2}\right) \qquad (n \to \infty), \tag{3.7}$$

for all points of $\{z_{k,n}\}_{k=1}^n \backslash C_\delta$. Again, because $\hat{z}_{k,n}$ is not necessarily the closest point of D_n to $z_{k,n}$, then dist $[z_{k,n}; D_n] \le \delta_{k,n}$ for all points of $\{z_{k,n}\}_{k=1}^n \backslash C_\delta$. Hence, the desired result of (1.9) of Theorem 4 follows.

To finally show that the result of (3.7) is *best possible* as a function of n, it again suffices to consider (as in §4.2) the special sequences $\{z_{m+1,2m+1}\}_{m=1}^\infty$ and $\{z_{m,2m}\}_{m=1}^\infty$ of zeros of $s_n(nz)$. Recalling the number μ of (2.21), it can be shown, in the manner of the proof of Theorem 3, that

$$\varliminf_{n\to\infty} \{n^2 \cdot \mathrm{dist}\ [\{z_{k,n}\}_{k=1}^n \backslash C_\delta; D_n]\} \ge \frac{\mu}{(1+\mu)^3} = 0.13326\cdots, \tag{3.8}$$

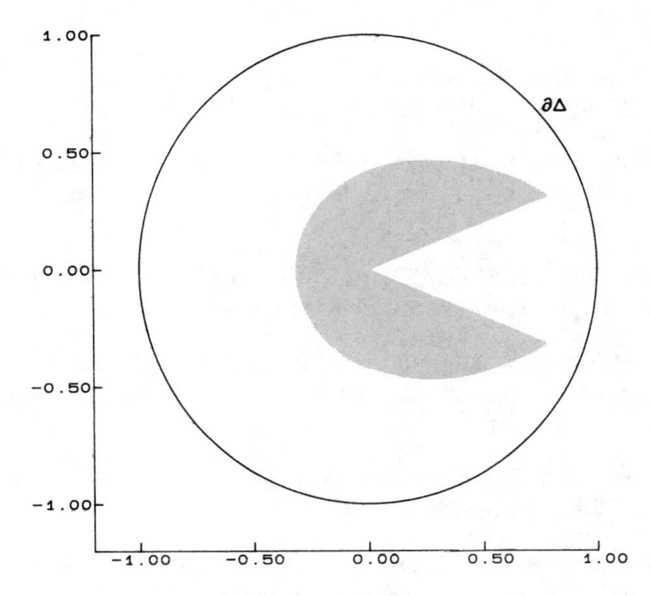

Figure 4.8: $\mathcal{D}_{27}$.

for *any* δ with $0 < \delta \leq 1$, which is the desired sharpness result of (1.9′). □

To conclude this chapter, consider the closed set

$$(3.9)\qquad \begin{aligned} \mathcal{D}_n := \Big\{ z \in \mathbb{C} : |ze^{1-z}|^n \leq \tau_n \sqrt{2\pi n}\ |1-z|/|z|, \quad |z| \leq 1 \text{ and} \\ |\arg z| \geq \cos^{-1}\left((n-2)/n\right) \Big\} \qquad (n = 1, 2, \cdots), \end{aligned}$$

whose boundary consists of the arc D_n of (1.7), and two line segments emanating from the origin. Figure 4.8 shows $\mathcal{D}_{27}$. We remark that the set $\mathcal{D}_n$ is evidently star-shaped with respect to the origin.

Because the zeros $\{z_{k,n}\}_{k=1}^{n}$ of $s_n(nz)$ are known to lie outside the Szegö curve $\mathcal{D}_\infty$, it is natural also to ask if the zeros of $s_n(nz)$ similarly lie *outside* of D_n for *every* $n \geq 1$. Unfortunately, there is not enough detail in Figures 4.4 and 4.5 to guess what the answer might be. Actually, based on high-precision calculations (with 120 significant figures) and some asymptotic analysis, the following two results can be stated:

$$(3.10)\qquad \begin{aligned} &(i) \text{ there exists a least positive integer } n_0 \text{ such that} \\ &\{z_{k,n}\}_{k=1}^{n} \bigcap \mathcal{D}_n \neq \phi \text{ for all positive integers } n > n_0; \text{ and} \\ &(ii)\ \{z_{k,n}\}_{k=1}^{n} \bigcap \mathcal{D}_n = \phi, \text{ for all } n = 1, 2, \cdots, 96. \end{aligned}$$

It further appears that $n_0 = 96$. These results can be found in [14].

REFERENCES

[1] N. Anderson, E.B. Saff, and R.S. Varga, *On the Eneström–Kakeya Theorem and its sharpness*, Linear Algebra Appl., **28** (1979), pp. 5-16.

[2] J.D. Buckholtz, *A characterization of the exponential series*, Amer. Math. Monthly, **73**, Part II (1966), pp. 121-123.

[3] A.J. Carpenter, *Some theoretical and computational aspects of approximation theory*, Ph.D. thesis, The University of Leeds, Leeds, England, 1988.

[4] A.J. Carpenter, R.S. Varga, and J. Waldvogel, *Asymptotics for the zeros of the partial sums of* e^z, I, Rocky Mountain J. Math., to appear.

[5] H.E. Fettis, J.C. Caslin, and K.R. Cramer, *Complex zeros of the error function and of the complementary error function*, Math. Comp., **27** (1973), pp. 401-404.

[6] P. Henrici, **Applied and Computational Complex Analysis**, Vol. 2, John Wiley & Sons, New York, 1977.

[7] M. Marden, **Geometry of Polynomials**, Mathematical Surveys No. 3, American Mathematical Society, Providence, RI, 1966.

[8] D.J. Newman and T.J. Rivlin, *The zeros of the partial sums of the exponential function*, J. Approx. Theory, **5** (1972), pp. 405-412.

[9] E.B. Saff and R.S. Varga, *On the zeros and poles of Padé approximants to* e^z, Numer. Math., **25** (1975), pp. 1-14.

[10] ———, *Zero-free parabolic regions for sequences of polynomials*, SIAM J. Math. Anal., **7** (1976), pp. 344-357.

[11] ———, *On the zeros and poles of Padé approximants to* e^z, III, Numer. Math., **30** (1978), pp. 241-266.

[12] G. Szegö, *Über eine Eigenschaft der Exponentialreihe*, Sitzungsber. Berl. Math. Ges., **23** (1924), pp. 50-64.

[13] E.C. Titchmarsh, **The Theory of Functions**, 2nd edition, Oxford University Press, London, 1950.

[14] R.S. Varga and A.J. Carpenter, *Asymptotics for the zeros of the partial sums of* e^z, II, **Computational Methods and Function Theory**, St. Ruscheweyh, E.B. Saff, L. Salinas, R.S. Varga, eds., Lecture Notes in Mathematics 1435, Springer–Verlag, Heidelberg, 1990, pp. 201-207.

CHAPTER 5

Real vs. Complex Best Rational Approximation

5.1. The numbers $\gamma_{m,n}$.

In this chapter, we describe a phenomenon, of recent research interest, which results when *complex* rational functions are pitted against *real* rational functions (of the same order) in approximating (in the uniform norm) *real* continuous functions on the *real* interval $[-1,+1]$.

For notation, let π_m^r and π_m^c be, respectively, the sets of polynomials (in the variable z or x) of degree at most m, with real and complex coefficients. For any pair (m,n) of nonnegative integers, $\pi_{m,n}^r$ and $\pi_{m,n}^c$ then denote, respectively, the sets of rational functions of the form p/q, with p in $\pi_m^r(\pi_m^c)$ and q in $\pi_n^r(\pi_n^c)$. With

$$I := [-1,+1],$$

let $C_r(I)$ be the set of all continuous real-valued functions on I. Then, for any f in $C_r(I)$, we further set

$$(1.1)\quad E_{m,n}^r(f) := \inf_{g\in\pi_{m,n}^r} \|f-g\|_{L_\infty(I)};\quad E_{m,n}^c(f) := \inf_{g\in\pi_{m,n}^c} \|f-g\|_{L_\infty(I)},$$

where, for any real- or complex-valued function h defined on I,

$$\|h\|_{L_\infty(I)} := \sup\{|h(x)| : x\in I\}.$$

The phenomenon to be studied in this chapter is this. We claim that, for each pair (m,n) of nonnegative integers with $n\geq 1$, there exists an f in $C_r(I)$ for which

$$(1.2)\quad \begin{cases} (i) & E_{m,n}^c(f) < E_{m,n}^r(f), \text{ and} \\ (ii) & \text{best uniform approximation to } f \text{ from } \pi_{m,n}^c \text{ on } I \\ & \text{is } \textit{not unique}. \end{cases}$$

Since $\pi_{m,n}^r$ is a proper subset of $\pi_{m,n}^c$ for each pair (m,n) of nonnegative integers, it is obvious from (1.1) that $E_{m,n}^c(f)\leq E_{m,n}^r(f)$ for any f in $C_r(I)$.

Because an arbitrary complex number consists of two real parameters, then any $r_{m,n}$ in $\pi^r_{m,n}$ can be associated with an $R_{m,n}$ in $\pi^c_{m,n}$ with *twice* the number of real parameters. This alone might heuristically suggest that $E^c_{m,n}(f)$ is roughly at most $E^r_{m,n}(f)/2$ in (1.2*i*) for *certain* f in $C_r(I)$. We shall later see in (1.19) that, except for the case $n = 0$ when $E^c_{m,0}(f) = E^r_{m,0}(f)$ for any f in $C_r(I)$ and any $m \geq 0$, this heuristically deduced inequality (and more) is essentially *correct*!

It is known (cf. Meinardus [5, p. 161]) that any f in $C_r(I)$ admits a *unique* best uniform approximant $r_{m,n}$ from $\pi^r_{m,n}$ on I. On the other hand, Walsh [14, p. 356] has given an example of a continuous complex-valued function, namely, $f(z) := z + z^{-1}$, which, on a certain compact crescent-shaped set in the complex plane, does *not* possess a unique best uniform rational approximation from $\pi^c_{1,1}$ on this set. That this phenomenon of *nonuniqueness* can hold even in the case of *real* functions on *real* intervals, as in (1.2*ii*), may come as somewhat of a surprise to the reader.

To give a concrete example exhibiting both parts of (1.2), we first recall (cf. [5, p. 161]) that, for any f in $C_r(I)$ and for any pair (m, n) of nonnegative integers, the unique $r_{m,n} = p/q$ in $\pi^r_{m,n}$ (where p and q are assumed to have no common factors) satisfying

$$E^r_{m,n}(f) = \|f - r_{m,n}\|_{L_\infty(I)}, \tag{1.3}$$

is precisely characterized by the existence of an *alternation set* $\{\xi_j\}_{j=1}^{\ell}$, with $-1 \leq \xi_1 < \xi_2 < \cdots < \xi_\ell \leq 1$, for which (with fixed $\lambda = 1$ or $\lambda = -1$)

$$f(\xi_j) - r_{m,n}(\xi_j) = \lambda(-1)^j E^r_{m,n}(f) \qquad (j = 1, 2, \cdots, \ell), \tag{1.4}$$

and for which

$$\ell \geq 2 + \max\{m + \deg q; n + \deg p\}. \tag{1.5}$$

(Here, we adopt the convention that if $p \equiv 0$, then $\deg p := -\infty$ and $\deg q := 0$, so that $\ell \geq 2 + m$ in (1.5) in this case. We also call ℓ the *length* of the alternation set $\{\xi_j\}_{j=1}^{\ell}$.)

Consider now the particular function $f(x) := x^2$ in $C_r(I)$. With $r_{1,1}(x) := p(x)/q(x) \equiv \frac{1}{2}$ for all real x, we have that

$$E^r_{1,1}(x^2) = \frac{1}{2}, \tag{1.6}$$

since all the conditions of (1.4) and (1.5) are fulfilled (with $\ell = 3$, $\lambda = -1$, $m = n = 1$, $\deg p = \deg q = 0$, $\xi_1 := -1$, $\xi_2 := 0$, and $\xi_3 := 1$). On the other hand, consider

$$g_{1,1}(x) := \frac{x + (\sqrt{2} - 1)i}{x + i}, \tag{1.7}$$

which is an element of $\pi^c_{1,1}$. A short calculation shows that

$$\|x^2 - g_{1,1}(x)\|_{L_\infty(I)} = \sqrt{2} - 1 = 0.41421 \cdots, \tag{1.8}$$

so that with (1.1), $E^c_{1,1}(x^2) \le 0.41421\cdots$. This implies from (1.6) that

$$E^c_{1,1}(x^2) < E^r_{1,1}(x^2). \tag{1.9}$$

Next, for any pair (m,n) of nonnegative integers, as well as for any f in $C_r(I)$, there always exists (cf. Walsh [14, p. 351]) an $R_{m,n}$ in $\pi^c_{m,n}$ for which $\|f - R_{m,n}\|_{L_\infty(I)} = E^c_{m,n}(f)$. Specifically, there is an $R_{1,1}$ in $\pi^c_{1,1}$ for which $\|x^2 - R_{1,1}\|_{L_\infty(I)} = E^c_{1,1}(x^2)$. But from (1.9), it is clear that $R_{1,1}(x)$ *cannot* be real for all real x. Consequently, as $x^2 - R_{1,1}(x)$ and its complex conjugate have the same uniform norm on I, then

$$E^c_{1,1}(x^2) = \|x^2 - R_{1,1}(x)\|_{L_\infty(I)} = \|x^2 - \overline{R_{1,1}(x)}\|_{L_\infty(I)}. \tag{1.10}$$

Thus, $R_{1,1}$ and $\overline{R_{1,1}}$, which are *distinct* elements in $\pi^c_{1,1}$, are *both* best uniform approximations to x^2 from $\pi^c_{1,1}$ on I, and the function x^2 exhibits both properties of (1.2) in the case $m = n = 1$. We remark that this argument shows in general that if $E^c_{m,n}(f) < E^r_{m,n}(f)$, as in (1.2*i*), then the nonuniqueness in (1.2*ii*) necessarily follows.

Gonchar first mentioned in 1968 this possibility of nonuniqueness in a footnote of his paper [2]. This possibility was followed up by Lungu, a student of Gonchar, who gave sufficient conditions in [4] in 1971 for the properties of (1.2) to hold. Independently, Saff and Varga [9], [10] made the same discovery in 1977, and obtained more general sufficient conditions for $E^c_{m,n}(f) < E^r_{m,n}(f)$ to hold for an f in $C_r(I)$, as well as a sufficient condition for $E^c_{m,n}(f) = E^r_{m,n}(f)$ to hold for an f in $C_r(I)$. The former sufficient conditions of Saff and Varga were later sharpened by Ruttan [6], who showed that $E^c_{m,n}(f) < E^r_{m,n}(f)$ holds if the best real uniform approximant from $\pi^r_{m,n}$ to f on I attains its maximum error on *no* alternation set (cf. (1.4)) of length greater than $m+n+1$, and that this lower bound is, in general, *best possible.* For a survey of such results, see [12, Chap. 5].

What we wish to focus on here is the following problem raised in Saff and Varga [10]. For each pair (m,n) of nonnegative integers, determine the nonnegative real number $\gamma_{m,n}$, defined by

$$\gamma_{m,n} := \inf\left\{E^c_{m,n}(f)/E^r_{m,n}(f) : f \in C_r(I)\backslash\pi^r_{m,n}\right\}. \tag{1.11}$$

In essence, determining the number $\gamma_{m,n}$ amounts to seeing just how much *better* best uniform approximation from $\pi^c_{m,n}$ on I can be, than best uniform approximation from $\pi^r_{m,n}$ on I, for *particular* functions in $C_r(I)\backslash\pi^r_{m,n}$. For example, for the function x^2 in $C_r(I)$, it is known (cf. Bennett, Rudnick, and Vaaler [1]) that $E^c_{1,1}(x^2) = (4/27)^{1/2} = 0.38490\cdots$, so that with (1.6),

$$\frac{E^c_{1,1}(x^2)}{E^r_{1,1}(x^2)} = 0.76980\cdots.$$

Consequently, this gives the following upper bound for $\gamma_{1,1}$:

$$\gamma_{1,1} \le 0.76980\cdots. \tag{1.12}$$

To precisely determine $\gamma_{m,0}$ for any nonnegative integer m, we first establish Proposition 1.

PROPOSITION 1. ([10]). *Given any f in $C_r(I)$ and given any pair (m,n) of nonnegative integers, then*

$$(1.13)\quad E^r_{m+n,2n}(f) \leq \inf_{g\in\pi^c_{m,n}} \|f - Re\ g\|_{L_\infty(I)} \leq E^c_{m,n}(f) \leq E^r_{m,n}(f).$$

Proof. As the final inequality of (1.13) is obvious, consider any p_m/q_n in $\pi^c_{m,n}$. On taking real parts, it is evident that

$$\left|f(x) - \frac{p_m(x)}{q_n(x)}\right| \geq \left|f(x) - Re\left(\frac{p_m(x)}{q_n(x)}\right)\right| \qquad (x \in I),$$

which establishes the second inequality of (1.13). Since $Re\ (p_m/q_n)$ is an element of $\pi^r_{m+n,2n}$, the first inequality of (1.13) then follows. □

On choosing $n = 0$ in (1.13), we see that

$$(1.14)\qquad E^c_{m,0}(f) = E^r_{m,0}(f) \qquad (f \in C_r(I); m = 0, 1, \cdots),$$

so that (cf. (1.11))

$$(1.15)\qquad \gamma_{m,0} = 1 \qquad (m = 0, 1, \cdots).$$

It turns out that the exact determination of the constants $\gamma_{m,n}$ of (1.11), when $m \geq 0$ and $n \geq 1$, is more delicate than the determination of $\gamma_{m,0}$ in (1.15). Four recent papers have described the behavior of $\gamma_{m,n}$ for $n \geq 1$. First, Trefethen and Gutknecht [11] established in 1983 the rather *remarkable* result that

$$(1.16)\qquad \gamma_{m,n} = 0 \qquad (n \geq m + 3; m = 0, 1, \cdots).$$

Next, Levin [3] established in 1986 the complementary result that

$$(1.17)\qquad \gamma_{m,n} = \frac{1}{2} \qquad (m + 1 \geq n \geq 1).$$

Levin's proof of (1.17) consisted of a direct construction to show that $\gamma_{m,n} \leq \frac{1}{2}$, and an algebraic method to show that $\gamma_{m,n} < \frac{1}{2}$ was impossible for $m+1 \geq n \geq 1$. The results of (1.16) and (1.17) left unresolved only the constants $\gamma_{m,m+2}$ $(m \geq 0)$. This case was most recently settled by Ruttan and Varga in 1989, where it was shown in [7] that $\gamma_{m,m+2} \leq \frac{1}{3}$ and in [8] that $\gamma_{m,m+2} < \frac{1}{3}$ was impossible. Thus,

$$(1.18)\qquad \gamma_{m,m+2} = \frac{1}{3} \qquad (m = 0, 1, \cdots).$$

The results of (1.15)–(1.18) give all the entries $\{\gamma_{m,n}\}_{m,n\geq 0}$ of Table 5.1.

TABLE 5.1
$\{\gamma_{m,n}\}_{m,n\geq 0}$.

n \ m	0	1	2	3	4	5	...
0	1	1	1	1	...	...	...
1	1/2	1/2	1/2	1/2	...	...	...
2	1/3	1/2	1/2	1/2	...	...	...
3	0	1/3	1/2	1/2	...	...	...
4	0	0	1/3	1/2	...	...	...
5	0	0	0	1/3	1/2	...	...
6	0	0	0	0	⋱	⋱	...
7	⋮	⋮	⋮	⋮	⋱	⋱	⋱
⋮	⋮	⋮	⋮	⋮	⋮	⋱	⋱

It is interesting that the constants $\{\gamma_{m,n}\}_{m,n\geq 0}$ of (1.11) take on only *four* distinct values: $1, \frac{1}{2}, \frac{1}{3}$, and 0, and only *three* distinct values when $n \geq 1 : \frac{1}{2}, \frac{1}{3}$, and 0. Specifically, this implies for any ε with $0 < \varepsilon < 1$ that, for each pair (m,n) of integers with $m \geq 0$ and $n \geq 1$, there exists an f in $C_r(I)\backslash\pi^r_{m,n}$ for which

$$(1.19) \quad \begin{cases} (i) \quad E^c_{m,n}(f) < (1+\varepsilon)E^r_{m,n}(f)/2 \qquad (m+1 \geq n \geq 1); \\ (ii) \quad E^c_{m,m+2}(f) < (1+\varepsilon)E^r_{m,m+2}(f)/3 \qquad (m = 0,1,\cdots); \\ (iii) \quad E^c_{m,n}(f) < \varepsilon E^r_{m,n}(f) \qquad (n \geq m+3; m = 0,1,\cdots), \end{cases}$$

which is a sharper form of (1.2*i*).

In the next sections, we give a unified treatment from [8] for the determination of the constants $\gamma_{m,n}$ when $m \geq 0$ and $n \geq 1$. We are also interested in those functions f in $C_r(I)\backslash\pi^r_{m,n}$ which satisfy

$$(1.20) \qquad \gamma_{m,n} < \frac{E^c_{m,n}(f)}{E^r_{m,n}(f)} < \gamma_{m,n} + \varepsilon \qquad (m \geq 0; n \geq 1),$$

for some given $\varepsilon > 0$. For such an f satisfying (1.20), if g in $\pi^c_{m,n}$ and if h in $\pi^r_{m,n}$ are such that

$$\|f-g\|_{L_\infty(I)} = E^c_{m,n}(f), \text{ and } \|f-h\|_{L_\infty(I)} = E^r_{m,n}(f),$$

we are likewise intrigued by the behavior of the associated errors:

$$f(x) - g(x) \text{ and } f(x) - h(x) \qquad (x \in I).$$

This will generate a number of interesting graphs in this chapter.

5.2. Upper bounds for $\gamma_{m,n}$.

As a means for obtaining upper bounds for $\gamma_{m,n}$ for $n \geq 1$, we establish the following proposition.

PROPOSITION 2. ([8]). *Given any pair (m, n) of nonnegative integers with $n \geq 1$, assume that*

$$\text{(2.1)} \qquad g \text{ is in } \pi^c_{m,n} \backslash \pi^r_{m,n}, \text{ with } Re\ g \text{ in } C_r(I),$$

and assume that S in $C_r(I)$ is such that there are $L \geq m+2$ distinct points $\{x_j\}_{j=1}^L$ with $-1 \leq x_1 < x_2 < \cdots < x_L \leq 1$, for which (with fixed $\lambda = 1$ or $\lambda = -1$)

$$\text{(2.2)} \qquad \lambda(-1)^j\left[S(x_j) + Re\ g(x_j)\right] > 0 \qquad (j = 1, 2, \cdots, L).$$

Then,

$$\text{(2.3)} \qquad \gamma_{m,n} \leq \|S - i\ Im\ g\|_{L_\infty(I)}/M,$$

where

$$\text{(2.4)} \qquad M := \min_{1 \leq j \leq L} |S(x_j) + Re\ g(x_j)|.$$

Proof. Set $f := S + Re\ g$, so that f is an element of $C_r(I)$. If the best uniform approximation to f in $\pi^r_{m,n}$ were the identically zero function, then the convention in (1.5) requires the existence of an alternation set in I of length $\ell \geq m+2$. Now, the hypothesis of (2.2) gives that f oscillates in sign in $L \geq m+2$ distinct points in I, and from this, using a result of de la Vallée-Poussin (cf. [5, p. 162]), one has the following lower bound for $E^r_{m,n}(f)$:

$$E^r_{m,n}(f) \geq M,$$

where M is defined in (2.4). On the other hand, as

$$E^c_{m,n}(f) \leq \|f - g\|_{L_\infty(I)} = \|S - i\ Im\ g\|_{L_\infty(I)},$$

we have from the definition of $\gamma_{m,n}$ in (1.11) that (2.3) is valid. □

It turns out that Trefethen and Gutknecht [11], Levin [3], and Ruttan and Varga [7] each, in essence, applied a variant of Proposition 2, with appropriate choices of $g(x)$ and $S(x)$, to determine upper bounds for $\gamma_{m,n}$. Their constructions of particular complex rational functions, which lead to sharp upper bounds for $\gamma_{m,n}$, are described in the next paragraphs.

We begin with the clever construction of a complex rational function, by Trefethen and Gutknecht [11], for establishing that $\gamma_{m,m+3} = 0$ $(m = 0, 1, \cdots)$. For any fixed nonnegative integer m and for any ε with $0 < \varepsilon < 1$ (and with

$0 < \varepsilon < 1/(2m-1)$ if $m \geq 1$), consider the following complex rational function in $\pi^c_{m,m+3} \backslash \pi^r_{m,m+3}$, defined by

$$h_{m,\varepsilon}(x) := \frac{\varepsilon \prod_{j=1}^{m} [-1 + (2j-1)\varepsilon - x]}{[x+1+\varepsilon]^{m+1}(i\sqrt{\varepsilon} - x)(1+\varepsilon - x)}, \tag{2.5}$$

(where, as usual, $\prod_{j=1}^{m} := 1$ if $m = 0$). It follows from (2.5) that

$$Re\ h_{m,\varepsilon}(x) = \frac{-\varepsilon x \prod_{j=1}^{m} [-1 + (2j-1)\varepsilon - x]}{[x+1+\varepsilon]^{m+1}(\varepsilon + x^2)(1+\varepsilon - x)} \in \pi^r_{m+1,m+4}, \tag{2.6}$$

and

$$Im\ h_{m,\varepsilon}(x) = \frac{-\varepsilon\sqrt{\varepsilon} \prod_{j=1}^{m} [-1 + (2j-1)\varepsilon - x]}{[x+1+\varepsilon]^{m+1}(\varepsilon + x^2)(1+\varepsilon - x)} \in \pi^r_{m,m+4}. \tag{2.7}$$

It is evident from (2.6) that $Re\ h_{m,\varepsilon}(x)$ has $m+1$ distinct zeros in $(-1, 0]$, with m closely packed zeros to the right of -1, plus an additional zero at the origin. Next, define the $m+2$ distinct points, namely, $\{x_j(\varepsilon) := -1 + 2j\varepsilon\}_{j=0}^{m}$ and $x_{m+1}(\varepsilon) := 1$, which satisfy

$$-1 = x_0(\varepsilon) < x_1(\varepsilon) < \cdots < x_{m+1}(\varepsilon) = 1.$$

These $m+2$ points $\{x_j(\varepsilon)\}_{j=0}^{m+1}$ *interlace* the $m+1$ zeros of $Re\ h_{m,\varepsilon}(x)$, and it can be verified that $Re\ h_{m,\varepsilon}(x)$ *oscillates* in sign in these points. Moreover, the pole of order $m+1$ at $-1-\varepsilon$ and the pole of order 1 at $1+\varepsilon$ of $Re\ h_{m,\varepsilon}(x)$ contribute in making these oscillations roughly of the same modulus, i.e., there is a constant c, dependent on m but independent of ε, such that (cf. [11])

$$(-1)^j Re\ h_{m,\varepsilon}(x_j(\varepsilon)) \geq c \| Im\ h_{m,\varepsilon} \|_{L_\infty(I)} / \sqrt{\varepsilon} \quad (j = 0, 1, \cdots, m+1). \tag{2.8}$$

With the rational function $h_{m,\varepsilon}(x)$ of (2.5) and with Proposition 2, we have the following theorem.

THEOREM 3. (Trefethen and Gutknecht [11]). *For any nonnegative integer* m *and for any integer* $n \geq m+3$,

$$\gamma_{m,n} = 0 \quad (n \geq m+3;\ m = 0, 1, \cdots). \tag{2.9}$$

Proof. We first show that $\gamma_{m,m+3} = 0$ for every nonnegative integer m. With $n := m+3$, set $S :\equiv 0, L := m+2$, and $g := h_{m,\varepsilon} \in \pi^c_{m,m+3}$, and apply Proposition 2. The discussion above shows that (2.2) of Proposition 2 is valid, and from (2.8) we see (cf. (2.4)) that

$$M := \min_{0 \leq j \leq m+1} |Re\ h_{m,\varepsilon}(x_j(\varepsilon))| \geq c \| Im\ h_{m,\varepsilon} \|_{L_\infty(I)} / \sqrt{\varepsilon}.$$

It thus follows from (2.3) of Proposition 2 that

$$\gamma_{m,m+3} \leq \sqrt{\varepsilon}/c. \tag{2.10}$$

But since c is independent of ε in (2.10) and since ε can be taken arbitrarily small, then

$$\gamma_{m,m+3} = 0. \tag{2.11}$$

Moreover, as $\pi^c_{m,n} \supset \pi^c_{m,m+3}$ for every $n \geq m+3$, the same function $h_{m,\varepsilon}$ can be used to deduce that

$$\gamma_{m,n} = 0 \quad (n \geq m+3; m = 0, 1, \cdots),$$

the desired result of (2.9). □

In the above construction, $f(x) := Re\ h_{m,\varepsilon}(x)$ of (2.6) is the function in $C_r(I)$ which is simultaneously approximated by the identically zero function of $\pi^r_{m,m+3}$ and by $h_{m,\varepsilon}(x)$ of $\pi^c_{m,m+3}$. Because of the oscillations of f in the $m+2$ distinct points of I of (2.8), the identically zero function in $\pi^r_{m,m+3}$ is a near best approximation to f from $\pi^r_{m,m+3}$ on I, and

$$E^r_{m,m+3}(f) \doteqdot \|f\|_{L_\infty(I)} = \|Re\ h_{m,\varepsilon}\|_{L_\infty(I)}.$$

Thus, in Figure 5.1 we graph the function $f(x) := Re\ h_{5,\varepsilon}(x)$ for x in I, with $\varepsilon = 0.1$, to show its seven oscillations in I. As some of these oscillations are very tiny, there is a 15-fold magnification (in y) given in Figure 5.2, of the dotted rectangular portion of Figure 5.1, which show these tiny oscillations more clearly. Next, the complex error in this case is, from our definitions, just $f(x) - g(x) = -i\ Im\ h_{5,\varepsilon}(x)$, and its path, as x increases from -1 to 1, in the complex plane is *uninteresting* (and omitted), since this path is just motions confined to a symmetric segment of the imaginary axis.

Concerning the result (2.9) of Theorem 3, it was pointed out by Saff that the *existence* of arbitrarily small numbers $\gamma_{m,n}$ was already implied in 1934 by a result of Walsh [13, Thm. IV], although this connection of Walsh's result to the numbers $\gamma_{m,n}$ was not previously noticed. Specifically, Walsh showed, for each fixed nonnegative integer m, that the functions $\bigcup_{n=0}^{\infty} \pi^c_{m,n}$ are *dense* in $C_c(I)$, the space of all continuous complex-valued functions on I. Thus, on choosing any f in $C_r(I)$, a subset of $C_c(I)$, this density implies that

$$\lim_{n\to\infty} E^c_{m,n}(f) = 0 \quad (m = 0, 1, \cdots). \tag{2.12}$$

On the other hand, let $T_j(x)$ $(j \geq 0)$ denote the jth Chebyshev polynomial (of the first kind). On specifically choosing $f(x) := T_{m+1}(x)$ in $C_r(I)$ and $r_{m,n} \equiv 0$ in $\pi^r_{m,n}$, then $f - 0$ has an alternation set in I of length exactly $m+2$, and it follows from the known fact that $\|T_{m+1}\|_{L_\infty(I)} = 1$ that (cf. (1.4))

$$E^r_{m,n}(f) = 1 \quad (n = 0, 1, \cdots). \tag{2.13}$$

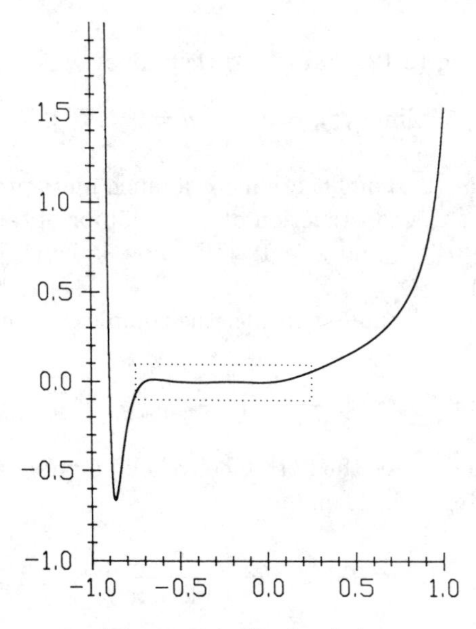

Figure 5.1: *Re* $h_{5,\varepsilon}(x)$.

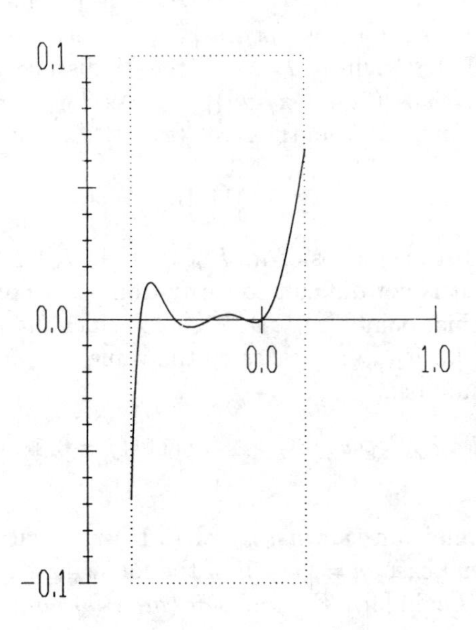

Figure 5.2: *Magnification.*

Obviously, combining (2.12) and (2.13) then gives

(2.14) $$\lim_{n\to\infty} \gamma_{m,n} = 0 \qquad (m = 0, 1, \cdots).$$

We see however that (2.9) of Theorem 3 is a much more precise form of (2.14).

We next present the construction of Levin [3] for obtaining upper bounds for $\gamma_{n+2k,n}$, where $n \geq 1$ and $k \geq 1$. (We show later in §3 that these upper bounds are *sharp*.)

For $\varepsilon > 0$ sufficiently small, consider the complex rational function defined by

(2.15) $$h_{2k,n,\varepsilon}(x) := T_{2k}(x)\left(\frac{x - i\varepsilon}{x + i\varepsilon}\right)^n \in \pi^c_{2k+n,n} \backslash \pi^r_{2k+n,n} \quad (n \geq 1),$$

where $T_{2k}(x)$ again denotes the $2k$th Chebyshev polynomial (of the first kind) for any positive integer k. Then,

(2.16) $$Re\ h_{2k,n,\varepsilon}(x) = T_{2k}(x)\ Re\left\{\left(\frac{x - i\varepsilon}{x + i\varepsilon}\right)^n\right\} \in \pi^r_{2n+2k,2n}.$$

For each real x, write $x + i\varepsilon = \rho e^{i\theta}$, so that $[(x - i\varepsilon)/(x + i\varepsilon)]^n = e^{-2ni\theta}$. This gives that

(2.17) $$Re\ \{[(x - i\varepsilon)/(x + i\varepsilon)]^n\} = \cos(2n\theta),$$

from which it follows that $Re\ \{[x - i\varepsilon)/(x + i\varepsilon)]^n\}$ has $2n$ distinct zeros, namely, $\{x_j(\varepsilon) := -\varepsilon \cot[(2j + 1)\pi/4n]\}_{j=0}^{2n-1}$, which are all clustered in an ε-neighborhood of the origin in I. Next, the $2k$ distinct zeros in I of $T_{2k}(x)$ are given by $\{y_i := \cos[(2i + 1)\pi/4k]\}_{i=0}^{2k-1}$. As $T_{2k}(0) = (-1)^k \neq 0$, these zeros of $T_{2k}(x)$ are *disjoint* from the zeros $\{x_j(\varepsilon)\}_{j=0}^{2n-1}$ for all $\varepsilon > 0$ sufficiently small, and hence

$$\{x_j(\varepsilon)\}_{j=0}^{2n-1} \bigcup \{y_i\}_{i=0}^{2k-1}$$

are the $2n + 2k$ (distinct) zeros of $Re\ h_{2k,n,\varepsilon}(x)$ in I, for all $\varepsilon > 0$ sufficiently small. From this, it is not difficult to verify from (2.16) and (2.17) that there are $2n+2k+1$ distinct points $\{w_j(\varepsilon)\}_{j=0}^{2n+2k}$ of I which *interlace* these zeros and for which (cf. [3]) $Re\ h_{2k,n,\varepsilon}(x)$ takes on the values $1 - o(1)$, with alternating sign, in these points, i.e.,

(2.18) $$(-1)^j\ Re\ h_{2k,n,\varepsilon}(w_j(\varepsilon)) \geq 1 - o(1) \quad (j = 0, 1, \cdots, 2n + 2k),$$

as $\varepsilon \to 0$.

With the rational function $h_{2k,n,\varepsilon}$ of (2.15) and with Proposition 2, we establish the special case $m = 2k + n$ of the following theorem.

THEOREM 4. (Levin [3]). *For any pair (m, n) of nonnegative integers with $m + 1 \geq n \geq 1$,*

(2.19) $$\gamma_{m,n} \leq \frac{1}{2} \qquad (m + 1 \geq n \geq 1).$$

Proof. We show that $\gamma_{2k+n,n} \leq \frac{1}{2}$. With $m := n + 2k$ where $k \geq 1$ and $n \geq 1$, set $S(x) := Re\ h_{2k,n,\varepsilon}(x)$, $L := 2n + 2k + 1$, and $g(x) := h_{2k,n,\varepsilon}(x) \in$

$\pi^c_{2k+n,n}$, and apply Proposition 2. First note that since $n \geq 1$ by hypothesis, then $L \geq m+2 = n+2k+2$. The discussion above shows that (2.2) of Proposition 2 is again valid, and from (2.18), we see (cf. (2.4)) that

$$M := \min_{0 \leq j \leq 2n+2k} |2 \; Re \; h_{2k,n,\varepsilon}(w_j(\varepsilon))| \geq 2 - o(1), \text{ as } \varepsilon \to 0.$$

Next, a calculation shows that

$$\|S(x) - i \; Im \; g(x)\|_{L_\infty(I)} = \|Re \; h_{2k,n,\varepsilon}(x) - i \; Im \; h_{2k,n,\varepsilon}(x)\|_{L_\infty(I)}$$

$$= \|\overline{h_{2k,n,\varepsilon}(x)}\|_{L_\infty(I)} = 1,$$

since, from (2.16), $\|(\frac{x+i\varepsilon}{x-i\varepsilon})^n\|_{L_\infty(I)} = 1 = \|T_{2k}\|_{L_\infty(I)}$ and $h_{2k,n,\varepsilon}(0) = 1$. Applying (2.3) of Proposition 2 then directly gives

$$\gamma_{2k+n,n} \leq \frac{1}{2-o(1)} \qquad (\varepsilon \to 0). \tag{2.20}$$

Letting $\varepsilon \to 0$ in the above expression results in

$$\gamma_{2k+n,n} \leq \frac{1}{2} \qquad (k = 1, 2, \cdots; n \geq 1),$$

the special case $m = 2k+n$ of (2.19). The construction for the remaining cases is similar (cf. [3]). □

In the previous construction, $f(x) = 2Re \; h_{2k,n,\varepsilon}(x)$ is a function in $C_r(I)$ which is simultaneously approximated by the identically zero function in $\pi^r_{2k+n,n}$ and by $h_{2k,n,\varepsilon}(x)$ in $\pi^c_{2k+n,n}$. In Figure 5.3, we graph the function $2 \; Re \; h_{2,5,\varepsilon}(x)$ for x in I with $\varepsilon = 0.1$, to show its nine points of near equioscillation in I. Next, the complex error in this case is just

$$f(x) - g(x) = \overline{h_{2,5,\varepsilon}(x)} \qquad (x \in I),$$

and its path, as x increases from -1 to 1, in the complex plane is given in Figure 5.4. Note the interesting *near-circularity* of this path!

Finally, we give the construction of Ruttan and Varga [7] for determining the upper bound $\gamma_{m,m+2} \leq \frac{1}{3}$ for any $m = 0, 1, \cdots$. For any ε satisfying $0 < \varepsilon < 1/(m+1)$, let m be any *fixed* nonnegative *even* integer and consider the following functions

$$\ell_j(z) := \ell_j(z; \varepsilon, m) := \frac{-\frac{2\varepsilon i}{3}(-1)^j}{z - 1 + \frac{2j}{m+1} - \varepsilon i} \qquad (j = 0, 1, \cdots, m+1). \tag{2.21}$$

It is evident from (2.21) that

$$\ell_j\left(1 - \frac{2j}{m+1}\right) = \frac{2}{3}(-1)^j, \text{ and } \ell_j\left(1 - \frac{2j}{m+1} \pm \varepsilon\right) = \frac{(1 \mp i)(-1)^j}{3}, \tag{2.22}$$

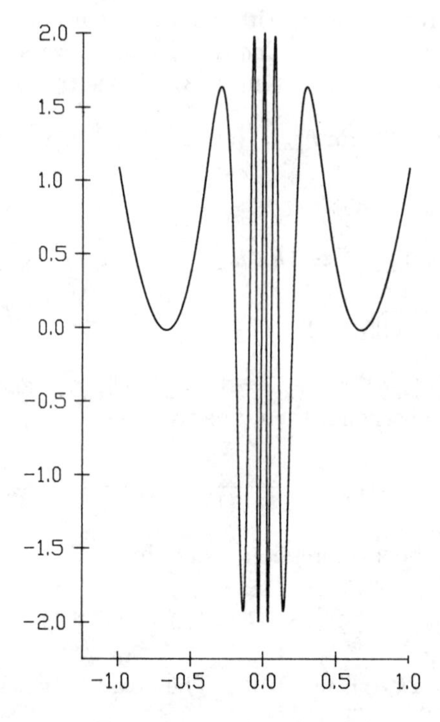

Figure 5.3: $2Re\ h_{2,5,\varepsilon}(x)$.

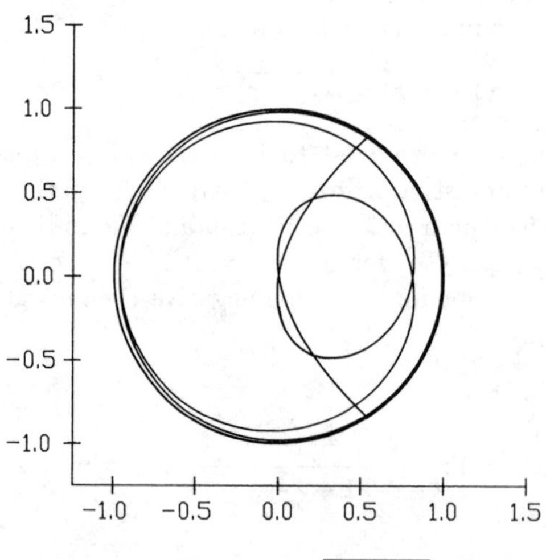

Figure 5.4: $\overline{h_{2,5,\varepsilon}(x)}$.

for all $j = 0, 1, \cdots, m+1$. Since each $\ell_j(z)$ is a Möbius transformation, each ℓ_j maps the real axis $\mathbb{R}$ onto some generalized circle in the complex plane. As $\ell_j(\infty) = 0$, this generalized circle necessarily passes through the origin. Moreover, as the pole of $\ell_j(z)$, namely, $1 - (2j)/(m+1) + \varepsilon i$, when reflected in $\mathbb{R}$, is the point $w_j := 1 - (2j)/(m+1) - \varepsilon i$, it follows from (2.21) that

$$\ell_j(w_j) = \frac{1}{3}(-1)^j \qquad (j = 0, 1, \cdots, m+1). \tag{2.23}$$

Thus, from the symmetry principle for Möbius transformations, the image of $\mathbb{R}$ under $w = \ell_j(z)$ is the circle with center $\frac{1}{3}(-1)^j$ and radius $\frac{1}{3}$ (since this circle passes through the origin). It is then geometrically clear that

$$\begin{gathered} \|Re\ \ell_j\|_{L_\infty(\mathbb{R})} = \|\ell_j\|_{L_\infty(\mathbb{R})} = \tfrac{2}{3}, \text{ and } \|Im\ \ell_j\|_{L_\infty(\mathbb{R})} = \tfrac{1}{3}, \\ \text{for } j = 0, 1, \cdots, m+1. \end{gathered} \tag{2.24}$$

To extend the results of (2.24), define the real intervals $I_k(m)$ by

$$I_k(m) := \left[1 - \frac{2k+1}{m+1}, 1 - \frac{2k-1}{m+1}\right] \bigcap I \qquad (k = 0, 1, \cdots, m+1), \tag{2.25}$$

so that these intervals cover $I := [-1, +1]$, i.e.,

$$\bigcup_{k=0}^{m+1} I_k(m) = I.$$

From the definitions of $\ell_j(x)$ and $I_k(m)$, it follows (as m is fixed) that

$$\|\ell_j\|_{L_\infty(I_k(m))} = O(\varepsilon) \text{ for any } k \neq j \qquad (\text{as } \varepsilon \to 0), \tag{2.26}$$

and from (2.22) that

$$\|Re\ \ell_j\|_{L_\infty(I_j(m))} = \frac{2}{3}, \text{ and } \|Im\ \ell_j\|_{L_\infty(I_j(m))} = \frac{1}{3} \ (j = 0, 1, \cdots, m+1). \tag{2.27}$$

Next, consider the complex rational function $g(x)$ defined by

$$h(x) = h(x; \varepsilon, m) := \sum_{j=0}^{m+1} \ell_j(x). \tag{2.28}$$

On rationalizing $h(x)$, we find that

$$h(x) = \frac{\frac{-2\varepsilon i}{3} \sum_{j=0}^{m+1} (-1)^j \prod_{\substack{k=0 \\ k \neq j}}^{m+1} \left\{x - 1 + \frac{2k}{m+1} - \varepsilon i\right\}}{\prod_{k=0}^{m+1} \left\{x - 1 + \frac{2k}{m+1} - \varepsilon i\right\}}, \tag{2.29}$$

so that h is at least an element of $\pi^c_{m+1,m+2}$. However, the numerator of $h(x)$ of (2.29) is

$$\frac{-2\varepsilon i}{3}\left\{x^{m+1}\sum_{j=0}^{m+1}(-1)^j + \text{ lower terms in } x^s(0 \le s \le m)\right\}.$$

But since m is by hypothesis even, then $\sum_{j=0}^{m+1}(-1)^j = 0$, which implies that h is an element of $\pi^c_{m,m+2}$. More precisely, it can be verified that the coefficient of x^m in the numerator of $h(x)$ of (2.29) is

$$-\frac{2(m+2)\varepsilon i}{3(m+1)} \neq 0,$$

so that h is an element of $\pi^c_{m,m+2}$, but not an element of $\pi^c_{s,m+2}$ for any $s < m$. It is interesting to mention here that (2.28) is just the *partial fraction decomposition* of $h(x)$.

With the rational function h of (2.29) and with Proposition 2, we have Theorem 5.

THEOREM 5. ([7]). *For each nonnegative integer* m,

$$\gamma_{m,m+2} \le \frac{1}{3}. \tag{2.30}$$

Proof. For a fixed nonnegative *even* integer m, consider the real continuous function $s(u)$ on $\mathbb{R}$, defined by

$$s(u) := \begin{cases} \dfrac{1-u^2}{1+u^2} & , \quad -1 \le u \le +1, \\ 0 & , \quad \text{otherwise}, \end{cases} \tag{2.31}$$

so that $s(0) = 1, s(\pm 1) = 0$, and $0 < s(u) < 1$ for $0 < |u| < 1$. Recalling that $0 < \varepsilon < 1/(m+1)$, set

$$S_{\varepsilon,m}(x) := \frac{1}{3}\sum_{j=0}^{m+1}(-1)^j s\left(\frac{x-1+\frac{2j}{m+1}}{\varepsilon^2}\right) \qquad (x \in \mathbb{R}). \tag{2.32}$$

It follows from (2.32) that $S_{\varepsilon,m}(x)$ is a real continuous function on $\mathbb{R}$ with

(2.33)

$$S_{\varepsilon,m}(1-\frac{2j}{m+1}) = \frac{(-1)^j}{3}, \text{ and } S_{\varepsilon,m}(1-\frac{2j}{m+1}\pm\varepsilon^2) = 0 \quad (j = 0, 1, \cdots, m+1).$$

Geometrically, we note that $S_{\varepsilon,m}(x)$ has $m+2$ alternating *spikes* on I, with one spike for each of the disjoint intervals $[1 - \frac{2j}{m+1} - \varepsilon^2, 1 - \frac{2j}{m+1} + \varepsilon^2]$ $(j = 0, 1, \cdots, m+1)$.

With the above definitions, set $n := m+2$, $L := m+2$, $S(x) := S_{\varepsilon,m}(x)$ of (2.32), and $g(x) := h(x;\varepsilon,m)$ of (2.28) and apply Proposition 2. With these definitions, we obtain from (2.22), (2.26)–(2.28), and (2.33) that

(2.34)
$$(-1)^j\left[S_{\varepsilon,m}\left(1-\frac{2j}{m+1}\right)+Re\ g\left(1-\frac{2j}{m+1}\right)\right]=1+O(\varepsilon)$$
$$(j=0,1,\cdots,m+1),$$

as $\varepsilon\to 0$, so that (cf. (2.4))

$$(2.35)\quad M:=\min_{0\le j\le m+1}\left|S_{\varepsilon,m}\left(1-\frac{2j}{m+1}\right)+\ Re\ g\left(1-\frac{2j}{m+1}\right)\right|=1+O(\varepsilon).$$

On the other hand, consider $S_{\varepsilon,m}(x)-i\ Im\ g(x)$ on I. For the particular interval $I_k(m)$ of (2.25), it follows from (2.26) that

$$S_{\varepsilon,m}(x)-i\ Im\ g(x)=S_{\varepsilon,m}(x)-i\ Im\ \ell_k(x)+O(\varepsilon)\qquad (x\in I_k(m)).$$

Moreover, a short calculation shows from (2.27) and (2.32) that

$$\|S_{\varepsilon,m}(x)-i\ Im\ \ell_k(x)\|_{L_\infty(I_k(m))}=\frac{1}{3}+O(\varepsilon)\qquad (k=0,1,\cdots,m+1),$$

so that with (2.26),

$$(2.36)\qquad \|S_{\varepsilon,m}(x)-i\ Im\ g(x)\|_{L_\infty(I)}=\frac{1}{3}+O(\varepsilon)\qquad (\text{as }\varepsilon\to 0).$$

Thus, it follows from (2.3) of Proposition 2 that

$$\gamma_{m,m+2}\le\frac{1}{3}+O(\varepsilon)\qquad (\text{as }\varepsilon\to 0),$$

and on letting $\varepsilon\to 0$, we have

$$\gamma_{m,m+2}\le\frac{1}{3},$$

the desired result of (2.30) when m is even.

For the case when m is odd, one simply modifies the definition of (2.21) to

$$(2.37)\quad \ell_j(z)=\ell_j(z;\varepsilon,m):=\frac{\frac{-2\varepsilon i}{3}\mu_j(-1)^j}{z-1+\frac{2j}{m+1}-\varepsilon\mu_j i}\qquad (j=0,1,\cdots,m+1),$$

where the numbers $\{\mu_j\}_{j=0}^{m+1}$ are any $m+2$ fixed positive numbers satisfying $0<\mu_j<1$ and

$$(2.38)\qquad \sum_{j=0}^{m+1}(-1)^j\mu_j=0\text{ and }\sum_{j=0}^{m+1}j(-1)^j\mu_j\ne 0.$$

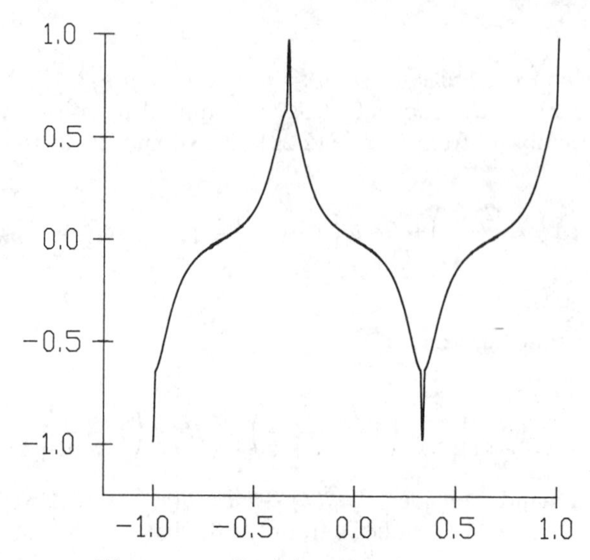

Figure 5.5: $S_{\varepsilon,2}(x) + Re\ h(x;\varepsilon,2)$.

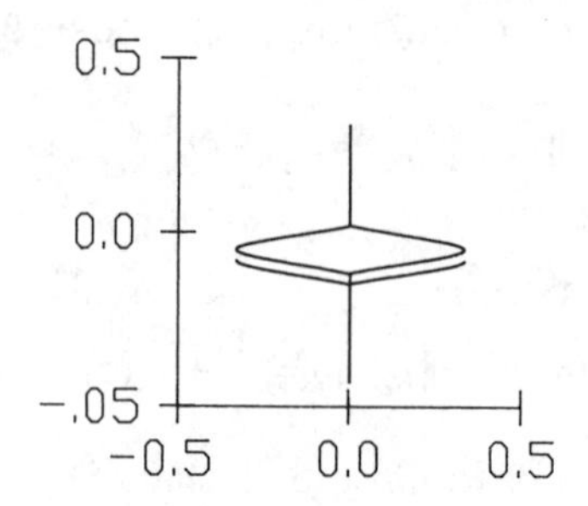

Figure 5.6: $S_{\varepsilon,2} - i\ Im\ h(x;\varepsilon,2)$.

An application of Proposition 2 again yields the desired result. □

In the above construction, $f_{\varepsilon,2m}(x) := S_{\varepsilon,2m}(x) + \ Re\ h(x;\varepsilon,2m)$ is the function in $C_r(I)$ which is simultaneously approximated by the identically zero function in $\pi^r_{2m,2m+2}$ and by $h(x;\varepsilon,2m)$ in $\pi^c_{2m,2m+2}$. In Figure 5.5, we graph the function $S_{\varepsilon,2}(x) + \ Re\ h(x;\varepsilon,2)$ for x in I with $\varepsilon = 0.1$, to show its four spikes and its four points of near equioscillation in I. Next, the complex error in this case is just $S_{\varepsilon,2}(x) - i\ Im\ h(x;\varepsilon,2)$, and its path, as x increases from -1 to 1, in the complex plane is given in Figure 5.6. This path is confined, from (2.36), to the disk $\{z \in \mathbb{C} : |z| \leq \frac{1}{3} + O(\varepsilon)\}$, but this path does not exhibit near-circularity.

The constructions and the figures of this section collectively show how *differently* the functions in $C_r(I)$ must be chosen in order to obtain sharp upper bounds for $\gamma_{m,n}$ in the three cases of (2.9), (2.19), and (2.30).

5.3. Lower bounds for $\gamma_{m,n}$.

To determine lower bounds for the $\gamma_{m,n}$, we describe the following two results of Ruttan and Varga [8, Thms. 4 and 5].

For a given real or complex polynomial p, let ∂p denote the *exact* degree of p. If $R = p/q$ is continuous on I where p and q are real polynomials, it is evident that $Re\ R = R$ can have at most ∂p sign changes in I, since each sign change of R corresponds to a zero of p. But, if $R = p/q$ is a continuous *complex-valued* function on I, the number of possible sign changes of $Re\ R$ on I depends not only on ∂p and ∂q, but also on the *magnitude* of the oscillations of $Re\ R$ in I. This is discussed in the following result of [8]. For additional notation, $\lfloor x \rfloor$ denotes the greatest integer N satisfying $N \le x$.

THEOREM 6. ([8]). *Let $\phi = p/q$ be a complex rational function, with no poles in I, which satisfies $\|Im\ \phi\|_{L_\infty(I)} \le 1$. Assume that there exist real numbers $d > 0$ and $\{x_j\}_{j=1}^{L}$, with $-1 \le x_1 < x_2 < \cdots < x_L \le 1$, for which (with fixed $\lambda = 1$ or $\lambda = -1$)*

$$\lambda(-1)^j\ Re\ \phi(x_j) \ge d \qquad (j = 1, 2, \cdots, L). \tag{3.1}$$

If $\partial q \le \partial p$ and if $d \ge 1$, then

$$L \le \partial p + 1. \tag{3.2}$$

Similarly, if $\partial q > \partial p$, then

$$L \le \partial q, \quad \text{whenever} \quad d \ge 1, \tag{3.3}$$

and

$$L \le \left\lfloor \frac{\partial p + \partial q + 1}{2} \right\rfloor, \quad \text{whenever} \quad d \ge 2. \tag{3.4}$$

Proof. We shall establish (3.2) using a geometrical argument, suggested by the work of Levin [3]. Assuming $\partial q \le \partial p$ and $d \ge 1$, let B denote the closed rectangle in the complex plane with vertices $\pm\ d \pm i$, so that the circle $C := \{z : |z| = 1\}$ is a subset of B, as indicated in Figure 5.7. Condition 3.1 and the hypothesis $\|Im\ \phi\|_L \le 1$ imply that the curve (in the extended complex plane) $\Gamma_1 := \{z = \phi(x) : x \in \mathbb{R}\}$ intersects the vertical sides of B, and hence the circle C, in $2(L-1)$ points as x increases from x_1 to x_L. (Here, points where Γ_1 is tangent to C are counted twice.) If $\phi(t)$, for t in I, is such an intersection of the Γ_1 and C, then

$$|\phi(t)|^2 = \left|\frac{p(t)}{q(t)}\right|^2 = 1,$$

so that t is a zero of the polynomial real polynomial

$$P(x) := |p(x)|^2 - |q(x)|^2.$$

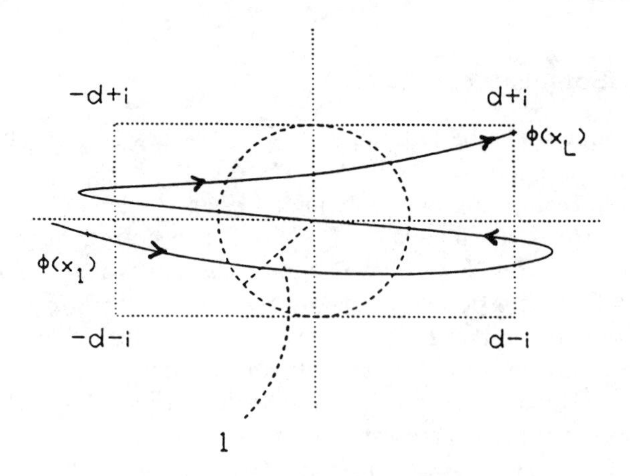

Figure 5.7: *Rectangle B.*

The above discussion shows that there are at least $2(L-1)$ zeros of $P(x)$ in I. Since $\partial p \geq \partial q$, then $\partial P \leq 2\partial p$. Thus, $2(L-1) \leq \partial P \leq 2\partial p$, which gives $L \leq \partial p + 1$, the desired result of (3.2). The proof of the remainder of Theorem 5 is similarly geometrical in nature. However, it involves many separate cases (cf. [8]). □

We remark, as shown in [8], that the results of (3.2)–(3.4) of Theorem 6 are *sharp*, in the following sense: (*i*) there exist complex rational functions, satisfying the appropriate hypotheses, for which the upper bounds for L given in (3.2)–(3.4) are attained, i.e., equality can hold in (3.2)–(3.4); and (*ii*) for any positive d with $d < 1 (< 2)$, there exist rational functions satisfying all hypotheses of Theorem 6 except the hypotheses on d in (3.3), (3.4), for which the bound on L in (3.3), (3.4) is *exceeded.*

With Theorem 6, the following lower bounds for $\gamma_{m,n}$ can be determined.

THEOREM 7. ([8]). *Let (m, n) be a pair of nonnegative integers with $n \geq 1$, let $f \in C^r(I) \backslash \pi^r_{m,n}$, and let $r_{m,n}$ and $R_{m,n}$ be, respectively, rational functions of best uniform approximation of f from $\pi^r_{m,n}$ and $\pi^c_{m,n}$ on I. Then,*

$$\frac{\|f - R_{m,n}\|_{L_\infty(I)}}{\|f - r_{m,n}\|_{L_\infty(I)}} > \frac{1}{2} \quad if \quad m+1 \geq n \geq 1, \tag{3.5}$$

and

$$\frac{\|f - R_{m,n}\|_{L_\infty(I)}}{\|f - r_{m,n}\|_{L_\infty(I)}} > \frac{1}{3} \quad if \quad m+2 \geq n \geq 1. \tag{3.6}$$

Consequently,

$$\gamma_{m,n} = \frac{1}{2} \quad if \quad m+1 \geq n \geq 1, \tag{3.7}$$

and

$$\gamma_{m,m+2} = \frac{1}{3}. \tag{3.8}$$

Proof. Let $s := \|f - R_{m,n}\|_{L_\infty(I)} / \|f - r_{m,n}\|_{L_\infty(I)}$, so that $0 \leq s \leq 1$, and set $e := f - r_{m,n}$, $R_{m,n} := p_1/q_1$, and $r_{m,n} := p_2/q_2$, where the pairs (p_1, q_1) and (p_2, q_2) are assumed to contain no common factors. Since $f \notin \pi^r_{m,n}$, we may assume, upon multiplication by a suitable nonzero constant, that $\|e\|_{L_\infty(I)} = 1$, so that

$$s = \|f - R_{m,n}\|_{L_\infty(I)}. \tag{3.9}$$

Since $r_{m,n}$ is the best uniform approximant of f from $\pi^r_{m,n}$ on I, there exist (cf. (1.5)) at least

$$L \geq 2 + \max[m + \partial q_2; n + \partial p_2] = m + m + 2 - \min[m - \partial p_2; n - \partial q_2] \tag{3.10}$$

distinct points $\{x_j\}_{j=1}^L$, with $-1 \leq x_1 < x_2 < \cdots < x_L \leq 1$, such that $e(x_j) = (-1)^j\lambda$ for all $1 \leq j \leq L$ (with fixed $\lambda = 1$ or $\lambda = -1$). Again, upon multiplication by -1 if necessary, we may assume that $\lambda = 1$, so that $e(x_1) = -1$.

With these normalizations, then

$$s = \|f - R_{m,n}\|_{L_\infty(I)} \geq |f(x_j) - R_{m,n}(x_j)| = |(-1)^j + r_{m,n}(x_j) - R_{m,n}(x_j)|$$

for all $1 \leq j \leq L$, which implies, on taking real parts, that

$$(-1)^j \{Re\ R_{m,n}(x_j) - r_{m,n}(x_j)\} \geq 1 - s \qquad (j = 1, 2, \cdots, L). \tag{3.11}$$

Next, set $\phi(x) := (R_{m,n}(x) - r_{m,n}(x))/s = p(x)/q(x)$, where p and q are complex polynomials with no common factors. From (3.11), it follows that

$$(-1)^j\ Re\ \phi(x_j) \geq (1 - s)/s =: d \qquad (j = 1, 2, \cdots L), \tag{3.12}$$

and since (cf. (3.9)) $s = \|f - R_{m,n}\|_{L_\infty(I)} \geq \|Im\ R_{m,n}\|_{L_\infty(I)}$, we similarly have

$$\|Im\ \phi\|_{L_\infty(I)} \leq 1. \tag{3.13}$$

Consider the sought result of (3.5) of Theorem 7, which, with our reductions and definitions, is the statement that

$$m + 1 \geq n \geq 1 \text{ implies } s > \frac{1}{2},$$

or equivalently, using the contrapositive, that $s \leq \frac{1}{2}$ implies $m + 1 < n$, i.e.,

$$s \leq \frac{1}{2} \text{ implies } m + 2 \leq n. \tag{3.14}$$

To establish (3.14), assume that s satisfies $0 < s \leq \frac{1}{2}$, which gives from (3.12) that $d \geq 1$. Because all the hypotheses of Theorem 6 are fulfilled, we see from (3.2) and (3.3) of Theorem 6 that

$$L \leq \partial p + 1 \text{ if } \partial q \leq \partial p, \text{ and } L \leq \partial q \text{ if } \partial q > \partial p. \tag{3.15}$$

Supposing that $\partial q > \partial p$, the second part of (3.15) gives $L \leq \partial q$, which can be expressed from (3.10) and the definitions $\phi = p/q = (R_{m,n} - r_{m,n})/s = (p_1/q_1 - p_2/q_2)/s$, as

$$m + m + 2 - \min[m - \partial p_2; n - \partial q_2] \leq \partial q = \partial q_1 + \partial q_2,$$

or equivalently, as

$$(3.16)\quad \{n - \partial q_1\} + \{(n - \partial q_2) - \min[m - \partial p_2; n - \partial q_2]\} \leq n - (m+2).$$

But as each term in braces of (3.16) is clearly nonnegative, then $n-(m+2) \geq 0$, giving $m + 2 \leq n$, the desired result of (3.14), under the assumption that $\partial q > \partial p$. However, as similar arguments show that the assumption $\partial q \leq \partial p$ leads to a contradiction, then (3.5) of Theorem 7 is valid. But then, it is evident from (3.5) that

$$\gamma_{m,n} \geq \frac{1}{2} \text{ if } m + 1 \geq n \geq 1,$$

and as the reverse inequality also holds from (2.19), then, as deduced in Levin [3],

$$\gamma_{m,n} = \frac{1}{2} \text{ if } m + 1 \geq n \geq 1,$$

the desired result of (3.7). Similar arguments (cf. [8]) establish (3.6), from which, with (3.20), the desired result of (3.8) follows. □

We note that the strict inequality in (3.5) gives us the *stronger* result that there is *no* f in $C_r(I)\backslash\pi^r_{m,n}$ for which

$$(3.17)\qquad \frac{\|f - R_{m,n}\|_{L_\infty(I)}}{\|f - r_{m,n}\|_{L_\infty(I)}} = \gamma_{m,n} = \frac{1}{2} \text{ if } m + 1 \geq n \geq 1,$$

where $r_{m,n} \in \pi^r_{m,n}$ and $R_{m,n} \in \pi^c_{m,n}$ are, respectively, the best uniform approximation from $\pi^r_{m,n}$ and $\pi^c_{m,n}$ of f on I. Thus, $\gamma_{m,n}$ is a *true infimum* (and not a minimum) as defined in (1.11), when $m + 1 \geq n \geq 1$. But the same is also true, for the same reasons, for $\gamma_{m,m+2} = \frac{1}{3}$ for all $m \geq 0$, and for $\gamma_{m,n} = 0$ for all $n \geq m + 3$. This is *why*, in retrospect, all the constructions of §5.2 depended on a small parameter $\varepsilon > 0$.

REFERENCES

[1] C. Bennett, K. Rudnick, and J.D. Vaaler, *Best uniform approximation by linear fractional transformations*, J. Approx. Theory, **25** (1979), pp. 204-224.

[2] A.A. Gonchar, *The rate of approximation by rational fractions and the properties of functions*, (in Russian), **Proceedings of the International Congress of Mathematicians**, Moscow, 1966, Izdat. "Mir," Moscow, 1968, pp. 329-356.

[3] A. Levin, *On the degree of complex rational approximation to real functions*, Constr. Approx., **2** (1986), pp. 213-219.

[4] K.N. Lungu, *Best approximation by rational functions*, (in Russian), Mat. Zametki, **10** (1971), pp. 11-15.

[5] G. Meinardus, **Approximation of Functions: Theory and Numerical Methods**, Springer-Verlag, Inc., New York, 1967.

[6] A. Ruttan, *The length of the alternation set as a factor in determining when a best real rational approximation is also a best complex rational approximation*, J. Approx. Theory, **31** (1981), pp. 230-243.

[7] A. Ruttan and R.S. Varga, *Real vs. complex rational Chebyshev approximation on an interval*: $\gamma_{m,m+2} \leq 1/3$, Rocky Mountain J. Math., **19** (1989), pp. 375-381.

[8] ———, *A unified theory for real vs. complex rational Chebyshev approximation on an interval* , Trans. Amer. Math. Soc., **312** (1989), pp. 681-697.

[9] E.B. Saff and R.S. Varga, *Nonuniqueness of best approximating complex rational functions*, Bull. Amer. Math. Soc., **83** (1977), pp. 375-377.

[10] ———, *Nonuniqueness of best complex rational approximations to real functions on real intervals*, J. Approx. Theory, **23** (1978), pp. 78-85.

[11] L.N. Trefethen and M.H. Gutknecht, *Real vs. complex rational Chebyshev approximation on an interval*, Trans. Amer. Math. Soc., **280** (1983), pp. 555-561.

[12] R.S. Varga, **Topics in Polynomial and Rational Interpolation and Approximation**, University of Montreal Press, Montreal, 1982.

[13] J.L. Walsh, *On approximation to an analytic function by rational functions of best approximations*, Math. Zeit., **38** (1934), 163-176.

[14] ———, **Interpolation and Approximation by Rational Functions in the Complex Domain**, 5th edition, Colloq. Publ. Vol. 20, Amer. Math. Soc., Providence, RI, 1969.

CHAPTER 6

Generalizations of Jensen's Inequality for Polynomials Having Concentration at Low Degrees

6.1. Polynomials having concentration at low degrees.

Recently, a new notion, i.e., for polynomials to have *concentration at low degrees*, has been introduced by Beauzamy and Enflo [2]. In this chapter, we present recent results on the application of this notion to generalizations of the classical Jensen's inequality in function theory. It is interesting to mention that many of the results to be discussed here were motivated (as in Beauzamy [3]) by numerical results using symbolic computer packages, such as MACSYMA.

To begin, let $p(z) = \sum_{j=0}^{m} a_j z^j$ denote any complex polynomial ($\not\equiv 0$). Given a real number d in $(0,1)$ and given a nonnegative integer k, then $p(z)$ is said to have *concentration* d *at degree* k *if*

$$\sum_{j=0}^{k} |a_j| \geq d \sum_{j=0}^{m} |a_j|. \tag{1.1}$$

(As we shall later see in §6.3, this concept extends to functions which are not polynomials.)

The first result established in this area was Theorem 1.

THEOREM 1. (Beauzamy and Enflo [2], [3]). *Given any real number d in $(0,1)$ and given any nonnegative integer k, let the real number $\tilde{C}_{d,k}$ (depending only on d and k) be defined as*

$$\tilde{C}_{d,k} := \sup_{1<t<\infty} \left[t \log \left\{ \frac{2d}{(t-1)[(\frac{t+1}{t-1})^{k+1} - 1]} \right\} \right]. \tag{1.2}$$

Then, for any polynomial $p(z) = \sum_{j=0}^{m} a_j z^j$ ($\not\equiv 0$) satisfying (1.1),

$$\frac{1}{2\pi} \int_0^{2\pi} \log |p(e^{i\theta})| d\theta - \log \left(\sum_{j=0}^{m} |a_j| \right) \geq \tilde{C}_{d,k}. \tag{1.3}$$

Remark. The important feature of Theorem 1 is that the lower bound, $\tilde{C}_{d,k}$, is *independent* of the degree of $p(z)$.

Proof. Let $p(z) = \sum_{j=0}^{m} a_j z^j$ be any polynomial ($\not\equiv 0$) satisfying (1.1). Without loss of generality, we normalize $p(z)$ by assuming that $\sum_{j=0}^{m} |a_j| = 1$. Thus from (1.3), it suffices to establish

$$\frac{1}{2\pi}\int_0^{2\pi} \log\left|p(e^{i\theta})\right| d\theta \geq \tilde{C}_{d,k}. \tag{1.3$'$}$$

For any r with $0 < r < 1$, Cauchy's formula gives

$$a_j = \frac{1}{2\pi}\int_0^{2\pi} \frac{p(re^{i\theta})d\theta}{r^j e^{ij\theta}} \qquad (j = 0, 1, \cdots, m),$$

from which it is evident, on taking absolute values, that

$$\sum_{j=0}^{k} |a_j| \leq \max_{|z|=r}\{|p(z)|\} \cdot \sum_{j=0}^{k} \frac{1}{r^j} = \max_{|z|=r} |p(z)| \cdot \left\{ \frac{\frac{1}{r^{k+1}} - 1}{\frac{1}{r} - 1} \right\}. \tag{1.4}$$

Then, choose any z_0 with $|z_0| = r$ such that $|p(z_0)| = \max_{|z|=r}|p(z)|$. Note that $|p(z_0)| > 0$ because $p(z) \not\equiv 0$.

Next, let $f(z)$ be any function analytic in $|z| \leq 1$ with $f(0) \neq 0$, and let $Z_\Delta(f)$ be the zeros of $f(z)$ in $0 \leq |z| < 1$. Then, *Jensen's formula* (cf. Ahlfors [1, p. 207]) is

$$\frac{1}{2\pi}\int_0^{2\pi} \log\left|f(e^{i\theta})\right| d\theta = \log|f(0)| + \sum_{z_j \in Z_\Delta(f)} \log\left(\frac{1}{|z_j|}\right). \tag{1.5}$$

Since the last term in (1.5) is nonnegative, one has from (1.5) the classical *Jensen's inequality*:

$$\frac{1}{2\pi}\int_0^{2\pi} \log\left|f(e^{i\theta})\right| d\theta \geq \log|f(0)|. \tag{1.5$'$}$$

Specifically, with the given polynomial $p(z)$ and the Möbius function $w(z) := (z + z_0)/(1 + \bar{z}_0 z)$ (which maps the disk $|z| \leq 1$ conformally onto the disk $|w| \leq 1$), set $f(z) := p(w(z)) = p\left(\frac{z+z_0}{1+\bar{z}_0 z}\right)$, which is analytic in $|z| \leq 1$ with $f(0) = p(z_0) \neq 0$. Then, Jensen's inequality (1.5$'$) applied to $f(z)$ directly gives

$$I := \frac{1}{2\pi}\int_0^{2\pi} \log\left|p\left(\frac{e^{i\theta} + z_0}{1 + \bar{z}_0 e^{i\theta}}\right)\right| d\theta \geq \log|p(z_0)|. \tag{1.6}$$

Next, with the change of variables $e^{i\phi} := \left(\frac{e^{i\theta}+z_0}{1+\bar{z}_0 e^{i\theta}}\right)$, a short calculation shows that

$$\begin{aligned} I &= \frac{1}{2\pi}\int_0^{2\pi} \log|p(e^{i\phi})| \cdot \frac{e^{i\phi}(1-r^2)}{\{-\bar{z}_0 e^{2i\phi} + (1+r^2)e^{i\phi} - z_0\}} d\phi \\ &= \frac{1}{2\pi}\int_0^{2\pi} \log|p(e^{i\phi})| \cdot \frac{1-r^2}{|1-\bar{z}_0 e^{i\phi}|^2} d\phi. \end{aligned} \tag{1.7}$$

For the second term in the integrand of the last integral in (1.7), we have the *lower bound*

$$\frac{1-r^2}{|1-\bar{z}_0 e^{i\phi}|^2} \geq \frac{1-r}{1+r} \qquad \text{(all real } \phi\text{)}, \tag{1.7$'$}$$

but, since $|p(e^{i\phi})| = \left|\sum_{j=0}^m a_j e^{ij\phi}\right| \leq \sum_{j=0}^m |a_j| = 1$, then the first term of this same integrand satisfies $\log|p(e^{i\phi})| \leq 0$. Thus, using the inequality of (1.7′), we have the following *upper bound* for I:

$$I \leq \frac{1}{2\pi}\left(\frac{1-r}{1+r}\right)\int_0^{2\pi} \log|p(e^{i\phi})|d\phi. \tag{1.8}$$

Combining the inequality (1.1) (under the normalization $\sum_{j=0}^m |a_j| = 1$) with (1.4), (1.6), and (1.8), gives

$$\frac{1}{2\pi}\int_0^{2\pi} \log|p(e^{i\phi})|d\phi \geq \left(\frac{1+r}{1-r}\right)\log\left[\frac{d(\frac{1}{r}-1)}{\frac{1}{r^{k+1}}-1}\right] \qquad (0<r<1),$$

and, with the change of variable $t := (1+r)/(1-r)$, this becomes (cf. (1.2))

$$\frac{1}{2\pi}\int_0^{2\pi} \log\left|p(e^{i\phi})\right| d\phi \geq t\log\left\{\frac{2d}{(t-1)[(\frac{t+1}{t-1})^{k+1}-1]}\right\} =: f_{d,k}(t), \tag{1.9}$$

for *any* t with $1 < t < \infty$. Clearly, the function $f_{d,k}(t)$ is real-valued and continuous on $(1, +\infty)$, and it is readily verified that

$$\tilde{C}_{d,k} := \sup_{1<t<\infty} f_{d,k}(t) \tag{1.10}$$

is *finite* for any d with $0 < d < 1$ and for any nonnegative integer k. Thus, since the inequality (1.9) holds for all t with $1 < t < \infty$, it also holds for $\tilde{C}_{d,k}$ of (1.10), which gives the desired result of (1.3′). □

As an interesting corollary to Theorem 1, consider the special case $k = 0$. From (1.9),

$$f_{d,0}(t) = t\log d \qquad (1<t<\infty),$$

so that

$$\tilde{C}_{d,0} := \sup_{1<t<\infty} f_{d,0}(t) = \log d.$$

Thus, for any polynomial $p(z) = \sum_{j=0}^m a_j z^j$ ($\not\equiv 0$) with $|a_0| = d\sum_{j=0}^m |a_j|$, (1.3) reduces in this case to

$$\frac{1}{2\pi}\int_0^{2\pi} \log|p(e^{i\theta})|d\theta \geq \log\left\{d\cdot\sum_{j=0}^m |a_j|\right\} = \log|a_0|,$$

which is just Jensen's inequality (1.5′). In this sense, the result (1.3) of Theorem 1 can be interpreted as a generalization of Jensen's inequality!

As remarked in Beauzamy [3], evaluating $f_{d,k}(t)$ at $t = 2$ gives (cf. (1.9)) the following *lower bound* for $\tilde{C}_{d,k}$:

$$\tilde{C}_{d,k} \geq 2 \log \left\{ \frac{2d}{3^{k+1} - 1} \right\}, \tag{1.11}$$

for all d with $0 < d < 1$ and for all nonnegative integers k. The proof of Theorem 1, as mentioned in [3], only makes use of the (weaker) hypotheses that $p(z) = \sum_{j=0}^{m} a_j z^j (\not\equiv 0)$ satisfies

$$\sum_{j=0}^{k} |a_j| \geq d \text{ and } \|p\|_{L^\infty(\Delta)} := \sup \{|p(z)| : |z| \leq 1\} \leq 1, \tag{1.12}$$

so that the conclusions of Theorem 1 hold under these assumptions.

Inequality (1.3) of Theorem 1 is an important one, and from it arises the obvious question of determining the *largest* real number $C_{d,k}$ for which (1.3) is valid for any polynomial $p(z)(\not\equiv 0)$ satisfying (1.1). Thus, if we introduce the functional

$$J(p) := \frac{1}{2\pi} \int_0^{2\pi} \log |p(e^{i\theta})| d\theta - \log \left(\sum_{j=0}^{m} |a_j| \right), \tag{1.13}$$

then

$$C_{d,k} := \inf \left\{ J(p) : p(z) = \sum_{j=0}^{m} a_j z^j (\not\equiv 0) \text{ satisfies (1.1)} \right\}. \tag{1.14}$$

Another obvious question is the characterization of *extremal polynomials*, i.e., those $p(z)$ satisfying (1.1) for which $C_{d,k} = J(p)$. This question remains *open*, in general, as of this writing.

It is illuminating to consider the special case $d = \frac{1}{2}$. For each positive integer k, the particular polynomial $(1 + z)^{2k+1}$ satisfies (with equality) (1.1) for $d = \frac{1}{2}$. Moreover, since $(1 + z)^{2k+1}$ has no zeros in $|z| < 1$, then Jensen's formula (1.5), with (1.13) and (1.14), gives

$$J\left((1+z)^{2k+1}\right) = -(2k+1)\log 2 \geq C_{1/2,k} \quad (k = 1, 2, \cdots). \tag{1.15}$$

As we shall see in the next section, the final inequality of (1.15) is in fact *sharp* for a large class of polynomials, for each positive integer k. There is, however, a considerable *gap* between the upper bound for $C_{1/2,k}$ in (1.15) and the lower bound for $\tilde{C}_{1/2,k}$ of Theorem 1. Specifically for the case $k = 1$, the upper bound (1.15) and the lower bound $\tilde{C}_{1/2,1}$ (determined numerically from (1.10)) for $C_{1/2,1}$ are

$$-2.07944\cdots \geq C_{1/2,1} \geq \tilde{C}_{1/2,1} = -3.69263\cdots. \tag{1.16}$$

6.2. Results for Hurwitz polynomials.

In this section, we describe new results of Rigler, Trimble, and Varga [12] on the Beauzamy–Enflo extension of Jensen's inequality, derived for the special set of *Hurwitz polynomials* H (cf. Marden [10, p. 181]), normalized for our purposes as

(2.1)

$$H := \begin{cases} f(z) = & \sum_{j=0}^{m} a_j z^j : m \text{ is an arbitrary nonnegative integer,} \\ & f(0) = 1, \text{ the coefficients } a_j \text{ are all real } (j = 0, 1, \cdots, m), \\ & \text{and all zeros of } f(z) \text{ lie in } Re\ z < 0. \end{cases}$$

Analogous to (1.14), for any d with $0 < d < 1$ and for any nonnegative integer k, we set

$$C_{d,k}^{H} := \inf\{J(f) : f(z) \in H \text{ and } f(z) \text{ satisfies } (1.1)\}, \tag{2.2}$$

where in general, from (1.13) and Jensen's formula (1.5),

$$J(f) := \log \left\{ \frac{|a_0|}{\sum_{j=0}^{m} |a_j| \cdot \prod_{z_j \in Z_\Delta(f)} |z_j|} \right\}. \tag{2.3}$$

Since H is a subset of *all* polynomials satisfying (1.1), it is evident from (1.14) and (2.2) that

$$C_{d,k}^{H} \geq C_{d,k}, \tag{2.4}$$

but, because of the special structure of elements in H, it turns out from the results of [12] that (*i*) all the constants $C_{d,k}^{H}$ can be *explicitly determined*, and that (*ii*) all the *extremal polynomials* in H, i.e., polynomials $q(z)$ in H for which

$$J(q) = C_{d,k}^{H}, \tag{2.5}$$

can also be found.

To begin, we first notice, as an immediate consequence of (2.1), that if $f(z) = \sum_{j=0}^{m} a_j z^j$ is in H, and if ∂f signifies the *exact* degree of $f(z)$, then

$$\begin{cases} (i) & a_0 = 1, \text{ and } a_j > 0 \text{ for all } j = 0, 1, \cdots, \partial f; \\ (ii) & \text{if } f(\rho) = 0 \text{ where } \rho < 0, \text{ then } f(z)/\left(1 - \frac{z}{\rho}\right) \text{ is in } H; \\ (iii) & \text{if } \rho < 0, \text{ then } f(z)\left(1 - \frac{z}{\rho}\right) \text{ is in } H; \\ (iv) & \text{if } f(\rho) = 0 \text{ where } \rho \text{ is nonreal,} \\ & \text{then } f(z)/[(1 - \frac{z}{\rho})(1 - \frac{z}{\bar{\rho}})] \text{ is in } H. \end{cases} \tag{2.6}$$

Next, if $f(z) = \sum_{j=0}^{m} a_j z^j$ is an element of H, then (2.6i) implies $f(1) = \sum_{j=0}^{m} a_j = \sum_{j=0}^{m} |a_j|$, and since $a_0 = 1$ from (2.6i), (2.3) reduces to

$$J(f) = -\log\left(f(1) \cdot \prod_{z_j \in Z_\Delta(f)} |z_j|\right) \qquad (f(z) \in H). \tag{2.7}$$

For convenience, define the following numbers for *any* $f(z) = \sum_{j=0}^{m} a_j z^j (\not\equiv 0)$:

$$\delta_k(f) := \sum_{j=0}^{k} |a_j| / \sum_{j=0}^{m} |a_j| \qquad (k = 0, 1, \cdots, m). \tag{2.8}$$

Note that if $f(z)$ has concentration d at degree k, then (cf. (1.1))

$$\delta_k(f) \geq d. \tag{2.9}$$

The following three lemmas from [12] are needed in the proof of our main result of this section, Theorem 5. We begin with the first of these results, Lemma 2. Its main contribution, equation (2.10), can be found in Mahler [8]. We remark that the discussion of the case of equality in (2.10) appears in [12]. (For related results, see Mahler [9].)

LEMMA 2. ([12] and Mahler [8]). *Consider any complex polynomial* $f(z) = \sum_{j=N}^{m} a_j z^j$ *where* $a_N a_m \neq 0$. *Then* (cf. (1.13)),

$$J(f) \geq -m \log 2, \tag{2.10}$$

with equality holding in (2.10) *only if* $N = 0$ *and* $f(z) = \gamma(e^{i\psi} + z)^m$, *where* $\gamma \neq 0$ *and* ψ *is real.*

Proof. From the general formula (2.3) for the functional $J(f)$, it is readily verified, for any polynomials $g(z)$ and $h(z)$ with $g(0) \cdot h(0) \neq 0$, that

$$J(gh) \geq J(g) + J(h).$$

Without loss of generality, we next consider any monic complex polynomial $f(z) = \sum_{j=0}^{m} a_j z^j = \prod_{j=1}^{m} (z + \zeta_j)$, with $|z_j| > 0 \quad (j = 1, 2, \cdots, m)$. Applying the above inductively extended inequality to $f(z)$ then yields

$$J(f) \geq \sum_{j=1}^{m} J(z + \zeta_j). \tag{2.10$'$}$$

Now, if $|z_j| \geq 1$, formula (2.3) gives that $J(z + z_j) = \log(|z_j|/(1 + |z_j|)) \geq -\log 2$, the last inequality following from the monotonicity of $u/(1+u)$ for $u \geq 1$. If $0 < |z_j| < 1$, (2.3) and similar reasoning gives that $J(z + z_j) = \log(1/(1 + |z_j|)) \geq -\log 2$. Applying these inequalities to (2.10$'$) gives

$$J(f) \geq -m \log 2,$$

the desired result of (2.10). The case of inequality in (2.10) is treated in [12]. $\square$

LEMMA 3. ([12]). *Let k be a positive integer, and consider any $f(z)$ in H with $\partial f := m \geq k+1$. Suppose that z_1 and z_2 are any two (not necessarily distinct) zeros of $f(z)$. Then, unless z_1 and z_2 are both real with $z_1 = -1$ and $z_2 \leq -1$ (or vice versa), there exists an $h(z)$ in H with $\partial h = m$ such that*

$$J(f) > J(h) \tag{2.11}$$

and (cf. (2.8))

$$\delta_k(h) > \delta_k(f). \tag{2.12}$$

Proof. First, suppose that at least one of $Im\ z_1$ and $Im\ z_2$ is not zero, say, $Im\ z_1 \neq 0$. The hypothesis that $f(z)$ is an element of H implies (cf. (2.1)) that $\bar{z}_1$ is also a zero of $f(z)$, and we *choose* $z_2 := \bar{z}_1$. Let the polynomials $g(z)$ and $h(z)$ be defined by

$$g(z) := \frac{f(z)}{\left(1 - \frac{z}{z_1}\right)\left(1 - \frac{z}{\bar{z}_1}\right)}, \text{ and } h(z) := \left(1 + \frac{z}{\rho}\right)^2 g(z), \tag{2.13}$$

where ρ satisfies $\rho > 1$. It is evident from (2.6) that $g(z)$ and $h(z)$ are both in H.

Since $\partial f = m$, we can write from (2.13) that $g(z) = \sum_{j=0}^{m-2} b_j z^j$, and since $g(z) \in H$ from (2.6*iv*) with $\partial g = m - 2$, we have (cf. (2.6*i*)) that $b_j > 0 \quad (j = 0, 1, \cdots, m-2)$. Note that the hypothesis $m \geq k+1$ implies $b_{k-1} > 0$. A short calculation then shows that

$$\delta_k(f) = \left\{ \sum_{j=0}^{k-1} b_j + \left[\left(|z_1|^2 b_k - b_{k-1} \right) / |1 - z_1|^2 \right] \right\} / g(1),$$

and

$$\delta_k(h) = \left\{ \sum_{j=0}^{k-1} b_j + \left[\left(\rho^2 b_k - b_{k-1} \right) / (1+\rho)^2 \right] \right\} / g(1).$$

Thus, $\delta_k(h) > \delta_k(f)$ iff

$$b_{k-1} \left[\frac{1}{|1 - z_1|^2} - \frac{1}{(1+\rho)^2} \right] > b_k \left[\frac{|z_1|^2}{|1 - z_1|^2} - \frac{\rho^2}{(1+\rho)^2} \right]. \tag{2.14}$$

With $Z_\Delta(f)$ again denoting the set of zeros of $f(z)$ with modulus less than 1, set $Z' := Z_\Delta(f) \setminus \{z_1; \bar{z}_1\}$. Then from (2.7) and (2.13),

$$J(f) = \log \left[\frac{\max\{|z_1|^2; 1\}}{g(1)|1 - z_1|^2 \prod_{\zeta \in Z'} |\zeta|} \right],$$

and

$$J(h) = \log \left[\frac{\rho^2}{g(1)(1+\rho)^2 \prod_{\zeta \in Z'} |\zeta|} \right].$$

Thus, $J(f) > J(h)$ iff

$$\frac{\max\{|z_1|; 1\}}{|1 - z_1|} > \frac{\rho}{1+\rho}. \tag{2.15}$$

If $|z_1| < 1$, then $\frac{1}{2} < 1/|1 - z_1| < 1$ because $Re\ z_1 < 0$. Hence, there is a $\rho > 1$ such that

$$\frac{1}{|1 - z_1|} > \frac{\rho}{1+\rho} > \frac{|z_1|}{|1 - z_1|}. \tag{2.16}$$

The first inequality of (2.16) shows that (2.15) is valid and, as $\rho > 1$, it also shows that $1/|1 - z_1| > 1/(1+\rho)$. Thus, the term multiplying b_{k-1} in (2.14) is *positive.* On the other hand, the last inequality of (2.16) shows that the term multiplying b_k in (2.14) is *negative.* As previously observed, $b_{k-1} > 0$, and from (2.6*i*), it follows that $b_k \geq 0$. Consequently, (2.14) is also valid, and (2.11) and (2.12) of Lemma 3 are thus established in this case.

If $|z_1| \geq 1$, then $\frac{1}{2} < (|z_1|/|1 - z_1|) < 1$ because $Re\ z_1 < 0$ and $z_1 \neq -1$. Hence, there is a $\rho_1 > 1$ such that

$$\frac{|z_1|}{|1 - z_1|} = \frac{\rho_1}{1 + \rho_1}.$$

Thus, $1 + 1/\rho_1 = |1 - 1/z_1| < 1 + 1/|z_1|$. This implies that $\rho_1 > |z_1|$ which, in turn, implies that $1/|1 - z_1| > 1/(1 + \rho_1)$. Hence, the right side of (2.14) is zero if $\rho = \rho_1$ and the left side is positive. It follows by continuity that there is some ρ in $(1, \rho_1)$ such that both (2.14) and (2.15) hold. Consequently, under the assumption that z_1 is a zero of $f(z)$ with $Im\ z_1 \neq 0$, there is an $h(z)$ in H which satisfies both (2.11) and (2.12) of Lemma 3.

Now, suppose that $Im\ z_1 = Im\ z_2 = 0$. There are three cases to be considered. First, assume that one of the numbers z_1 and z_2 is in the open interval $(-1, 0)$, say, $-1 < z_1 < 0$. Redefine $g(z)$ and $h(z)$ by

$$g(z) := \frac{f(z)}{\left(1 - \frac{z}{z_1}\right)} \qquad \left(\text{and we write } g(z) = \sum_{j=0}^{m-1} b_j z^j \right),$$

and

$$h(z) := (1 + z)g(z).$$

A similar calculation to that preceding the derivation of (2.14) shows that

$$\delta_k(f) = \left[\sum_{j=0}^{k-1} b_j - z_1 b_k/(1 - z_1) \right] /g(1),$$

and

$$\delta_k(h) = \left[\sum_{j=0}^{k-1} b_j + b_k/2\right] / g(1).$$

In this case, we also have $b_k > 0$. Thus, $\delta_k(h) > \delta_k(f)$ iff $\frac{1}{2} > -z_1/(1-z_1)$, which is always true for z_1 in the interval $(-1, 0)$. Redefine $Z' := Z_\Delta(f)\backslash\{z_1\}$. From (2.7) and the new definitions of $g(z)$ and $h(z)$, we have

$$J(f) = \log\left[\frac{1}{g(1)(1-z_1)\prod_{\zeta\in Z'}|\zeta|}\right],$$

and

$$J(h) = \log\left[\frac{1}{2g(1)\prod_{\zeta\in Z'}|\zeta|}\right].$$

Thus, $J(f) > J(h)$ iff $1/(1-z_1) > \frac{1}{2}$, and this last inequality is certainly true. This completes the first case.

Next, suppose that $Im\ z_1 = Im\ z_2 = 0$ and assume that both z_1 and z_2 are in the interval $(-\infty, -1)$. Additionally, assume that

$$1 - z_1 - z_2 - z_1 z_2 \geq 0. \tag{2.17}$$

Redefine $g(z)$ and $h(z)$ by

$$g(z) := \frac{f(z)}{(1-\frac{z}{z_1})(1-\frac{z}{z_2})} \qquad \left(\text{and we write } g(z) = \sum_{j=0}^{m-2} b_j z^j\right),$$

and

$$h(z) := (1+z)\left(1+\frac{z}{\rho}\right)g(z), \text{ where } \rho > 1.$$

As in the derivation of (2.14), $\delta_k(h) > \delta_k(f)$ iff

$$\begin{aligned} &b_{k-1}\left[\frac{1}{(1-z_1)(1-z_2)} - \frac{1}{2(1+\rho)}\right] \\ &> b_k\left[\frac{z_1 z_2}{(1-z_1)(1-z_2)} - \frac{\rho}{2(1+\rho)}\right]. \end{aligned} \tag{2.18}$$

As in the derivation of (2.15), $J(f) > J(h)$ iff

$$\frac{z_1 z_2}{(1-z_1)(1-z_2)} > \frac{\rho}{2(1+\rho)}. \tag{2.19}$$

If equality holds in (2.17), then the left side of (2.19) becomes equal to $\frac{1}{2}$, and (2.19) is true for all $\rho \geq 1$. Further, since the term multiplying b_k in (2.18) is positive and tends to zero as $\rho \to \infty$ and since in this case we similarly have that $b_{k-1} > 0$, it follows that the inequality in (2.18) can be made to be valid by choosing $\rho \geq 1$ sufficiently large.

Next, assume strict inequality holds in (2.17). Coupling this with the fact that z_1 and z_2 are in $(-\infty, -1)$, we have that $\frac{1}{4} < z_1 z_2/[(1-z_1)(1-z_2)] < \frac{1}{2}$. Consequently, there is a $\rho_2 > 1$ such that

$$\frac{z_1 z_2}{(1-z_1)(1-z_2)} = \frac{\rho_2}{2(1+\rho_2)}.$$

In turn, this implies that

$$\frac{1}{2(1+\rho_2)} = \frac{2-(z_1+1)(z_2+1)}{2(1-z_1)(1-z_2)} < \frac{1}{(1-z_1)(1-z_2)}.$$

Thus, the right side of (2.18) is zero if $\rho = \rho_2$ and, since $b_{k-1} > 0$ as before, the left side of (2.18) is positive when $\rho = \rho_2$. It follows by continuity that there is some ρ in the interval $(1, \rho_2)$ such that both (2.18) and (2.19) hold. Again, this implies that there is an $h(z)$ in H which satisfies both (2.11) and (2.12) of Lemma 3.

Finally, suppose that $Im\ z_1 = Im\ z_2 = 0$, assume that z_1 and z_2 are both in $(-\infty, -1)$, but that (2.17) does not hold. With $g(z)$ as previously defined, redefine $h(z)$ by $h(z) := (1 + z/\rho)g(z)$, where $\rho > 1$. Then, $\delta_k(h) > \delta_k(f)$ iff

$$\text{(2.20)} \qquad \frac{b_{k-1}}{(1-z_1)(1-z_2)} > b_k \left[\frac{z_1 z_2}{(1-z_1)(1-z_2)} - \frac{\rho}{1+\rho}\right],$$

and $J(f) > J(h)$ iff

$$\text{(2.21)} \qquad \frac{z_1 z_2}{(1-z_1)(1-z_2)} > \frac{\rho}{1+\rho}.$$

Since (2.17) does *not* hold, then $\frac{1}{2} < z_1 z_2 / [(1-z_1)(1-z_2)] < 1$. Thus, there is a $\rho_3 > 1$ such that

$$\frac{z_1 z_2}{(1-z_1)(1-z_2)} = \frac{\rho_3}{1+\rho_3}.$$

By continuity, there is some ρ in $(1, \rho_3)$ such that both (2.20) and (2.21) hold; whence, there is an $h(z)$ in H which satisfies both (2.11) and (2.12) of Lemma 3. □

It is interesting to point out that the motivation for Lemma 3 in [12] also came from computer experiments using MACSYMA, where, for a polynomial $p(z)$ in H satisfying the concentration constraint (1.1), its complex zeros were moved in *pairs* to reduce the functional $J(p)$ of (1.13), while still satisfying the constraint (1.1).

LEMMA 4. ([12]). *Given any d in $(0,1)$ and given any nonnegative integer k, there is a unique positive integer n, dependent on d and k, such that*

$$\frac{1}{2^n}\sum_{j=0}^{k}\binom{n}{j} \le d < \frac{1}{2^{n-1}}\sum_{j=0}^{k}\binom{n-1}{j}. \tag{2.22}$$

Moreover, if the number $\hat{\rho}$ is defined by

$$\hat{\rho} := \frac{\binom{n-1}{k}}{\sum_{j=0}^{k}\binom{n-1}{j} - d\,2^{n-1}} - 1, \tag{2.23}$$

then $\hat{\rho} \ge 1$.

Proof. Given any nonnegative integer k, consider the sequence of positive real numbers

$$\left\{\frac{1}{2^\ell}\sum_{j=0}^{k}\binom{\ell}{j}\right\}_{\ell=k}^{\infty}, \tag{2.24}$$

whose initial term is unity. We claim that this sequence is strictly decreasing and has limit zero. To see this, set

$$a_\ell := \frac{1}{2^\ell}\sum_{j=0}^{k}\binom{\ell}{j} \qquad (\ell = k, k+1, \cdots). \tag{2.25}$$

Since

$$\binom{\ell+1}{j} = \binom{\ell}{j} + \binom{\ell}{j-1} \qquad (j = 1, 2, \cdots, \ell), \tag{2.26}$$

if follows from (2.25) that

$$a_{\ell+1} = a_\ell - \frac{1}{2^{\ell+1}}\binom{\ell}{k} \qquad (\ell = k, k+1, \cdots),$$

which implies that the terms of (2.24) are strictly decreasing. Next, as a nice consequence of the *Central Limit Theorem* (cf. Patel and Read [11, pp. 169–170]), we have that

$$\left| a_\ell - \frac{1}{\sqrt{2\pi}}\int_{-\infty}^{(2k+1-\ell)/\sqrt{\ell}} e^{-t^2/2}dt \right| < \frac{0.28}{\sqrt{\ell}} \tag{2.27}$$

for all $\ell \ge \max\{k; 1\}$. As k is fixed, (2.27) shows that $a_\ell \to 0$ as $\ell \to +\infty$. (For an easier proof that $a_\ell \to 0$ as $\ell \to \infty$, note that (2.25) directly gives, for all $\ell \ge 2k$, that

$$0 < a_\ell < \frac{(k+1)\ell^k}{k!2^\ell},$$

and, since k is fixed, this upper bound clearly tends to zero as $\ell \to \infty$.)

Thus, given any d in $(0,1)$, the strictly decreasing nature of the a_ℓ of (2.25) implies there is a unique positive integer n with $n \geq k+1$, such that (2.22) is satisfied. It follows directly from (2.22) and (2.26) that $\hat{\rho}$, defined in (2.23), satisfies $\hat{\rho} \geq 1$. □

This brings us now to the main result of this section.

THEOREM 5. ([12]). *Given any d in $(0,1)$ and given any positive integer k, let n and $\hat{\rho}$ be (uniquely) defined from* (2.22) *and* (2.23). *Then* (cf. (2.2)),

$$C_{d,k}^{H} = \log\left[\frac{\hat{\rho}}{(\hat{\rho}+1)2^{n-1}}\right]. \tag{2.28}$$

Moreover, set $Q_{n,\hat{\rho}}(z) := (1+\frac{z}{\hat{\rho}})(1+z)^{n-1}$, which is an element of H. Then, $f(z)$ in H satisfying (1.1) *is an extremal polynomial in H* (i.e., $J(f) = C_{d,k}^{H}$) *iff $f(z) = Q_{n,\hat{\rho}}(z)$.*

Proof. Given a d in $(0,1)$ and given a positive integer k, let the positive integer n and real number $\hat{\rho}$ with $\hat{\rho} \geq 1$ be the uniquely defined quantities from (2.22) and (2.23) of Lemma 4. Then, the polynomial $Q_{n,\hat{\rho}}(z) := (1+(z/\hat{\rho}))\,(1+z)^{n-1}$ is an element of H which, as can be verified, satisfies

$$\delta_k(Q_{n,\hat{\rho}}) = d \text{ and } J(Q_{n,\hat{\rho}}) = \log\left[\frac{\hat{\rho}}{(\hat{\rho}+1)2^{n-1}}\right]. \tag{2.29}$$

By definition (cf. (2.2)), (2.29) implies that

$$J(Q_{n,\hat{\rho}}) = \log\left[\frac{\hat{\rho}}{(\hat{\rho}+1)2^{n-1}}\right] \geq C_{d,k}^{H}. \tag{2.30}$$

Now, as $\log\left[\rho/(\rho+1)2^{n-1}\right]$ is strictly increasing as a function of ρ for $\rho \geq 1$ and is bounded above by $-(n-1)\log 2$, it follows from Lemma 2 that there is no need to consider polynomials in H of degree less than n. It follows (cf. (2.24)) from the definition of n in (2.22), that $n \geq k+1$. Lemma 3 then implies that it is sufficient to consider any $f(z) = (1+\frac{z}{\rho})(1+z)^{m-1}$, where $m \geq n$ and $\rho \geq 1$. Since this $f(z)$ must satisfy (1.1), it can be shown that $m \leq n$, and if $m = n$, then $\hat{\rho} \leq \rho$, where $\hat{\rho}$ is as defined in (2.23). Thus, it is only necessary to consider the case when $m = n$ and $\hat{\rho} \leq \rho$. A computation based on (2.7) shows that

$$J((1+\frac{z}{\rho})(1+z)^{n-1}) = \log\left\{\frac{\rho}{(1+\rho)2^{n-1}}\right\}. \tag{2.31}$$

But the quantity on the right in (2.31) is again strictly increasing for $\rho \geq \hat{\rho}$. Hence, to minimize the above quantity, we must choose $\rho = \hat{\rho}$, which gives not only that

$$\inf\{J(f) : f(z) \in H \text{ and } f(z) \text{ satisfies } (1.1)\} = J(Q_{n,\hat{\rho}}) = C_{d,k}^{H}, \tag{2.32}$$

but also that $Q_{n,\hat{\rho}}(z)$ is the *unique* element in H with $J(Q_{n,\hat{\rho}}) = C^H_{d,k}$. □

To conclude this section, we mention that considerable numerical experimentation has led us to believe that equality holds in (2.4), and we make the

$$\textbf{Conjecture } ([12]).\ C^H_{d,k} \stackrel{?}{=} C_{d,k}, \quad \textit{for all } 0 < d < 1 \textit{ and all } k = 0, 1, \cdots, \tag{2.33}$$

for all d in $(0, 1)$ and all nonnegative integers k.

6.3. Generalizations and remarks.

In light of the fact that the results of §§6.1 and 6.2, derived for polynomials, do not restrict the *degrees* of the polynomials considered, it is not surprising that these results carry over *without change* to certain entire functions *of order zero* (cf. Boas [5]). As shown in [12], the results of Theorem 5 are in fact valid for the elements of the set

$$\tilde{H} := \left\{ f(z) = \prod_{j=1}^{\omega} \left(1 - \frac{z}{z_j}\right) (\text{ with } \omega \leq \infty) : z_k \in \{z_j\}_{j=1}^{\omega} \right.$$
$$\left. \text{implies both } Re\, z_k < 0 \text{ and } \bar{z}_k \in \{z_j\}_{j=1}^{\omega}, \text{ and } \sum_{j=1}^{\omega} \frac{1}{|z_j|} < \infty \right\}.$$

Note that $\tilde{H}$ contains the set H of (2.1) as a *proper subset.*

In another direction, to go beyond functions analytic in the *closed* unit disk $|z| \leq 1$, let $f(z) = \sum_{j=0}^{\infty} a_j z^j$ be analytic in $|z| < 1$, and set

$$M_\infty(r; f) := \max_{0 \leq \theta \leq 2\pi} |f(re^{i\theta})|, \text{ for each } r \text{ with } 0 \leq r < 1. \tag{3.1}$$

As usual, the Hardy space H^∞ is defined (cf. Duren [7, p. 2]) as

$$H^\infty := \{g(z) : g(z) \text{ is analytic in } |z| < 1 \text{ and } M_\infty(r; g) \text{ is bounded as } r \to 1-\}. \tag{3.2}$$

Suppose, for an $f(z) = \sum_{j=0}^{\infty} a_j z^j$ which is analytic in $|z| < 1$, we consider the appropriateness of the concept of (1.1) for $f(z)$ having concentration d at degree k. For this to be meaningful, we must have $\sum_{j=0}^{\infty} |a_j| < \infty$, so that (cf. (3.1))

$$M_\infty(r; f) \leq \sum_{j=0}^{\infty} |a_j| r^j \leq \sum_{j=0}^{\infty} |a_j| < \infty;$$

whence from (3.2), $f(z) \in H^\infty$. It turns out (cf. [7, p. 17]) that any $f(z)$ in H^∞ can be extended to the boundary $|z| = 1$ by means of a function $\hat{f}(e^{i\theta})$, defined on $[0, 2\pi]$, for which

$$(3.3)\qquad \begin{cases} \hat{f}(e^{i\theta}) = \lim_{r\to 1-} f(re^{i\theta}) \text{ almost everywhere in } [0, 2\pi]; \\ \hat{f}(e^{i\theta}) \in L^\infty[0, 2\pi]; \\ \text{if } f(z) \not\equiv 0, \text{ then } \log|\hat{f}(e^{e\theta})| \in L^1[0, 2\pi]. \end{cases}$$

Thus, for any $f(z) = \sum_{j=0}^\infty a_j z^j$ ($\not\equiv 0$) which is analytic in $|z| < 1$ with $\sum_{j=0}^\infty |a_j| < \infty$, then $f(z) \in H^\infty$, and the functional analogue of (1.13), namely,

$$(3.4)\qquad \hat{J}(f) := \frac{1}{2\pi}\int_0^{2\pi} \log|\hat{f}(e^{i\theta})|d\theta - \log\left(\sum_{j=0}^\infty |a_j|\right),$$

is then well defined and finite. Accordingly, one has the associated problem of minimizing $\hat{J}(f)$ in (3.4) over all such $f(z)$ in H^∞ satisfying $\sum_{j=0}^k |a_j| \geq d\sum_{j=0}^\infty |a_j|$. For the case when d satisfies $0 < d < 1$ and when $k = 0$, this minimization (cf. [12, Thm. 1]) takes place iff $|a_0| = d\sum_{j=0}^\infty |a_j|$ and $f(z)$ is its own associated *outer function* (cf. Rudin [13, p. 338]).

The concept of a polynomial $p(z) = \sum_{j=0}^m a_j z^j$ having concentration d at degree k, i.e., (cf. (1.1))

$$(3.5)\qquad \sum_{j=0}^k |a_j| \geq d\sum_{j=0}^m |a_j|,$$

can be easily generalized, since what is involved in the above equation is simply the ℓ_1-norm of the vector of coefficients $(a_0, a_1, \cdots, a_m)^T$, and the ℓ_1-norm of its *k-restriction*, defined as $(a_0, a_1, \cdots, a_k, 0, \cdots, 0)^T$. Clearly, one can instead consider ℓ_q-norms in (3.5), i.e.,

$$(3.6)\qquad \left(\sum_{j=0}^k |a_j|^q\right)^{1/q} \geq d\left(\sum_{j=0}^m |a_j|^q\right)^{1/q} \qquad (1 \leq q \leq \infty),$$

and its associated minimization problems, or one can even mix norms, such as (cf. (1.12))

$$(3.7)\qquad \left(\sum_{j=0}^k |a_j|^q\right)^{1/q} \geq d\|p(z)\|_{L^\infty(\Delta)} \qquad (1 \leq q \leq \infty),$$

for other associated minimization problems. Such generalizations have been considered in Beauzamy [4] and Bonvalot [6].

Next, given d with $0 < d < 1$ and given a nonnegative integer k, we remark that there are polynomials $p(z)$, *not* in H, which satisfy both (1.1) and

$$(3.8)\qquad J(p) = C_{d,k}^H.$$

As a simple example of this, consider, as in Theorem 5, the particular cubic polynomial

$$Q_{n,1}(z) := (1+z)^3 = 1 + 3z + 3z^2 + z^3, \tag{3.9}$$

which is an element of $\mathcal{H}$ and which satisfies (1.1) (with equality) for $d = \frac{1}{2}$ and $k = 1$. From (2.28) of Theorem 5, it follows that

$$J(Q_{n,1}) = -3\log 2 = C^H_{1/2,1}. \tag{3.10}$$

On the other hand, consider the quintic polynomial

$$p_5(z) := (1+z)^3(4-z^2) = 4 + 12z + 11z^2 + z^3 - 3z^4 - z^5, \tag{3.11}$$

which is not an element of H of (2.1), as $p_5(z)$ has a zero in $Re\ z > 0$. Now, $p_5(z)$ also satisfies (1.1) (with equality) for $d = \frac{1}{2}$ and $k = 1$, and, in addition, on using (2.3), $p_5(z)$ similarly satisfies

$$J(p_5) = -3\log 2 = C^H_{1/2,1}. \tag{3.12}$$

This implies, of course, that if the conjecture of (2.33) is valid, i.e., $C^H_{d,k} = C_{d,k}$, then extremal polynomials $p(z)$ satisfying $J(p) = C_{d,k}$ are *not* necessarily uniquely determined, in sharp contrast with the result of Theorem 5!

Finally, to illustrate how scientific computations with MACSYMA can be applied in this chapter, we include Figures 6.1 and 6.2 and the following discussion. Specifically, take the polynomial $p_5(z)$ of (3.11) and vary its single zero, +2, while simultaneously satisfying the constraint (1.1) (with equality) for $d = \frac{1}{2}$ and $k = 1$. Thus, we consider the polynomial

$$P_\mu(z) := \{(1+z)^3(2+z)\}\,[\mu - z] \tag{3.13}$$

where μ is a complex parameter with initial value +2, and where $P_\mu(z)$ is subject to the constraint (cf. (2.8)) that $\delta_1(P_\mu) = \frac{1}{2}$. Then, it can be shown that there exists a simple closed curve Γ in the complex plane, passing through $\mu = 2$, where Γ is defined by

$$\Gamma := \left\{\mu \in \mathbb{C} : \delta_1(\mathrm{P}_\mu) = \frac{1}{2}\right\}. \tag{3.14}$$

Because the multiplier, $h(z) := \{(1+z)^3(2+z)\}$, of $[\mu - z]$ in (3.13), satisfies $\delta_1(h) = \frac{3}{8} < \frac{1}{2}$, we note that this curve Γ cannot extend to infinity.

In Figure 6.1, we have plotted the curve Γ, and it can be verified that

$$\min\{J(P_\mu) : \mu \in \Gamma\} = J(P_2) = -3\log 2; \tag{3.15}$$

more precisely, the minimum value of $J(P_\mu)$ on Γ occurs only when $\mu = 2$. Thus, as we traverse the curve Γ, we generate a sequence of polynomials $P_\mu(z)$ which all satisfy $\delta_1(P_\mu) = \frac{1}{2}$, but no improvement, over that of (3.12), is obtained in minimizing the associated functional J over this sequence of polynomials.

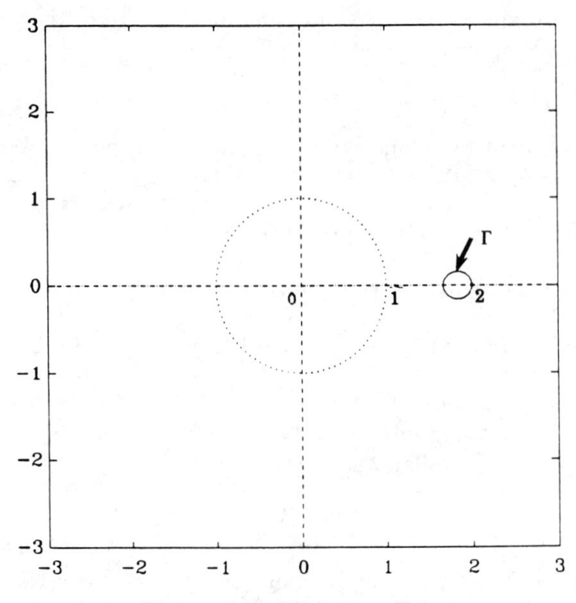

Figure 6.1: *The curve* Γ.

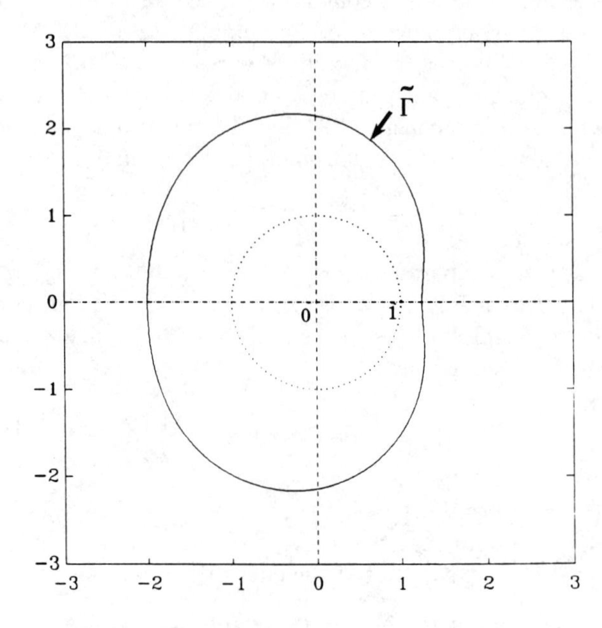

Figure 6.2: *The curve* $\tilde{\Gamma}$.

We can also take the polynomial $p_5(z)$ of (3.11) and vary its single zero, -2, while simultaneously satisfying the constant $\delta_1(\tilde{P}_\mu) = \frac{1}{2}$, where

$$\tilde{P}_{\tilde{\mu}}(z) := \left\{(1+z)^3(2-z)\right\}[-\tilde{\mu}+z],$$

and where $\tilde{\mu}$ is a complex parameter with initial value -2. This gives rise to the simple closed curve

$$\tilde{\Gamma} := \left\{\tilde{\mu} \in \mathbb{C} : \delta_1(\tilde{\mathrm{P}}_{\tilde{\mu}}) = \frac{1}{2}\right\},$$

passing through -2. This is shown in Figure 6.2. As in (3.15), it is again the case that

$$\min\left\{J(\tilde{P}_{\tilde{\mu}}) : \tilde{\mu} \in \tilde{\Gamma}\right\} = J(\tilde{P}_{-2}) = -3\log 2. \tag{3.16}$$

REFERENCES

[1] L.V. Ahlfors, **Complex Analysis**, 3rd edition, McGraw-Hill Book Co., New York, 1979.

[2] B. Beauzamy and P. Enflo, *Estimations de produits de polynômes*, J. Number Theory, **21** (1985), pp. 390-412.

[3] B. Beauzamy, *Jensen's inequality for polynomials with concentration at low degrees*, Numer. Math., **49** (1986), pp. 221-225.

[4] ———, *A minimization problem connected with the generalized Jensen's inequality*, J. Math. Anal. Appl., **145**(1990), pp. 137-144.

[5] R.P. Boas, **Entire Functions**, Academic Press, Inc., New York, 1954.

[6] L. Bonvalot, *Moyene géométrique des fonctions des espaces de Hardy et polynômes concentrés aux bas degrés*, Thèse de Troisième Cycle, Université de Paris 7, Paris, 1986.

[7] P.L. Duren, **Theory of H^p Spaces**, Academic Press, Inc., New York, 1970.

[8] K. Mahler, *An application of Jensen's formula to polynomials*, Mathematika, **7** (1960), pp. 98-100.

[9] ———, *On two extremum properties of polynomials*, Ill. J. Math., **7** (1963), pp. 681-701.

[10] M. Marden, **Geometry of Polynomials**, Mathematical Surveys Number 3, American Mathematical Society, Providence, RI, 1966.

[11] J.K. Patel and C.B. Read, **Handbook of the Normal Distribution**, Marcel Dekker, New York, 1982.

[12] A.K. Rigler, S.Y. Trimble, and R.S. Varga, *Sharp lower bounds for a generalized Jensen inequality*, Rocky Mountain J. Math., **19** (1989), pp. 353-373.

[13] W. Rudin, **Real and Complex Analysis**, McGraw-Hill Book Co., New York, 1966.

Index